D0772781

Innovation with Chinese Characteristics

Innovation with Chinese Characteristics

High-Tech Research in China

Edited by

Linda Jakobson
Finnish Institute of International Affairs (FIIA)

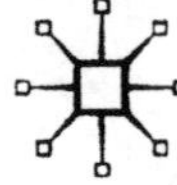

First published in 2007 by
PALGRAVE MACMILLAN
Houndmills, Basingstoke, Hampshire RG21 6XS and
175 Fifth Avenue, New York, N.Y. 10010
Companies and representatives throughout the world.

PALGRAVE MACMILLAN is the global academic imprint of the Palgrave Macmillan division of St. Martin's Press, LLC and of Palgrave Macmillan Ltd. Macmillan® is a registered trademark in the United States, United Kingdom and other countries. Palgrave is a registered trademark in the European Union and other countries.

ISBN-13: 978–0–230–00692–8 hardback
ISBN-10: 0–230–00692–2 hardback

This book is printed on paper suitable for recycling and made from fully managed and sustained forest sources. Logging, pulping and manufacturing processes are expected to conform to the environmental regulations of the country of origin.

A catalogue record for this book is available from the British Library.

A catalog record for this book is available from the Library of Congress.

10 9 8 7 6 5 4 3 2 1
16 15 14 13 12 11 10 09 08 07

Printed and bound in Great Britain by
Antony Rowe Ltd, Chippenham and Eastbourne

Contents

Notes on Contributors

Chunli Bai is Executive Vice President of the Chinese Academy of Sciences (CAS). He serves as the chief scientist for the National Steering Committee for Nanoscience and Related Technology and Director of the National Center for Nanoscience and Technology of China. In addition, he is the President of the Graduate School of CAS, a Fellow of the Academy of Sciences for the Developing World (TWAS), and a Foreign Associate of National Academy of Sciences (U.S.). Bai graduated from the Department of Chemistry, Beijing University in 1978 and received his MS and PhD degrees from CAS. He worked at Caltech in the U.S. for two years in the early 1980s. In the mid-1980s, he shifted his research orientation to scanning tunneling microscopy and molecular nanotechnology. Bai has won numerous awards in China and abroad.

Linda Jakobson is the Beijing-based Director of the China Program at the Finnish Institute of International Affairs (FIIA). She has written five books on China and East Asia during the 14 years she has lived in China and other parts of East Asia. The Finnish edition of *A Million Truths. A Decade in China* (New York: M. Evans, 1998) won the Finnish Government Publication Award. In 1990, Jakobson was a Fellow at the Kennedy School of Government at Harvard University. A Mandarin speaker, she has also written about the Taiwan Straits and grassroots political reform in China, among others, "Greater Chinese Union" in *The Washington Quarterly* (2005), and a chapter about village elections in *Governance in China* (Jude Howell ed., Rowman & Littlefield, 2004).

Kejun Jiang is Research Professor and Director of Energy System Analysis and Market Analysis Division at the Energy Research Institute under the National Development and Research Commission (NDRC). He received his PhD in Social Engineering at the Tokyo Institute of Technology in 1999. Since 1993, Jiang has been doing climate change related energy policy analysis, which focuses on energy technology and energy supply, renewable energy development, and energy conservation. Jiang's recent studies include energy and emission scenarios, assessment of energy and fuel tax, and energy targets in China.

Arthur Kroeber is co-editor of the *China Economic Quarterly* (CEQ) and director of its parent company, Dragonomics Research & Advisory, an independent research firm specializing in analysis of the Chinese economy and its global impact. He graduated from Harvard in 1984 and subsequently worked as a journalist and economic analyst in India and China before joining the CEQ in 2002. He has published articles in the *Economist* and the *Far Eastern Economic Review* and is a frequent contributor to the opinion page of the *Financial Times*.

Chen Wang is the Deputy Director of the National Center for Nanoscience and Technology of China. Prior to that, he was the Director of the Laboratory for Molecular Nanostructures and Nanotechnology at the Institute of Chemistry (CAS). Wang received his BS from the University of Science and Technology of China in 1986 and his PhD in physics from the University of Virginia in 1992. He is an adjunct Professor at the Department of Chemistry in Beijing University, Jilin University, and Central China Normal University. In 2002–03, he was a Visiting Scholar at the University of California at Berkeley and Lawrence Berkeley Laboratory. Wang has worked on the applications of scanning probe microscopy in surface characterizations and fabrications. He has published over 100 peer-reviewed papers in international journals.

Jun Yu is Professor and Associate Director of Beijing Institute of Genomics of CAS. Prior to this he worked at New York University (1990–93) and University of Washington Genome Center (1993–1998). Yu obtained his BS in biochemistry from Jilin University in 1983 and PhD in biomedical sciences from New York University Medical School in 1990. His primary research interests include genome biology and bioinformatics. He has led many major genome projects in China and has published over 100 scientific papers and over a dozen books and book chapters. Yu has won numerous academic awards both in China and abroad.

List of Figures

Foreword: Innovation in China

Antti Hautamäki

China has already for a long time been the "manufacturer of the world," producing everything from toys to cars and computers. Almost all major companies from the European Union, the United States and Japan have established production plants in China. Chinese exports are strongly linked to exports by China-based foreign or foreign-Chinese joint venture companies. Until now, product development and innovation have mostly taken place in the research and development (R&D) labs of companies and universities of industrial countries. But the situation is changing rapidly. We can even talk about a new phase of globalization, in which developing countries like China and India, not to say anything about such well developed societies as South Korea and Taiwan, are taking a more active role also in R&D and innovation. Although Chinese universities and research labs seem to imitate Western science, it's only a question of time before they will be capable of producing high quality research and breakthrough innovation. It is noteworthy that that by establishing joint R&D-driven companies with Chinese partners, Western and Japanese firms transfer know-how to Chinese firms and professionals.

From now on, the "China phenomena" will change the knowledge balance of the world. All Western countries, including Finland, are carefully following the development of China's innovation system and innovation policy. The development of the Chinese innovation environment is going to challenge the leading edge status of many Western innovation hubs. But globalization is not only competition, it is also global collaboration between most advanced regions and knowledge hubs, and between firms and universities, wherever they might be. From this point of view, keeping abreast with the development of the Chinese innovation system will open new perspectives to cooperation with Chinese universities and companies.

These are some of the reasons that propelled Sitra to ask researcher Linda Jakobson of the Finnish Institute of International Affairs (FIIA) to take responsibility for a book about innovation and high-tech research in China. The plan was to produce five focused chapters by Beijing-based specialists, to transmit a view from China. Linda Jakobson has written the first chapter, an overview painting a broader picture about innovation and R&D policies in China. The following chapters, which Jakobson describes in the Introduction, focus on high-tech research in the fields of information technology (IT), nanotech, energy, and biotech. They have been written by Arthur Kroeber, Chunli Bai and Chen Wang, Kejun Jiang, and Jun Yu, respectively. All these fields are interesting to Western countries; they are also the Chinese government's major areas of investment and research. Energy is at the same time a global political issue related to both global warming and the growing demand of energy in China.

The resulting book *Innovation with Chinese Characteristics* is a compact presentation of the ongoing transformation of innovation in the most important fields of technology. The book gives new insight that is valuable to a wide array of specialists and companies—anyone seeking to better understand the Chinese innovation markets. On behalf of Sitra, The Finnish Innovation Fund, I would like to thank Linda Jakobson for her success in the impossible task to edit a short book about innovation and high-tech research in China.

In Berkeley, California, 15 January 2007

Antti Hautamäki

Director, Innovations and New Solutions
Sitra, the Finnish Innovation Fund

Abbreviations

ANGCC	advanced natural gas combined cycle
APPCDC	Asia Pacific Partnership for Clean Development and Climate (sometimes also abbreviated as AP6)
BGI	Beijing Genomics Institute
BIG	Beijing Institute of Genomics
CAE	Chinese Academy of Engineering
CAS	Chinese Academy of Sciences
CAST	Chinese Association for Science and Technology
CCID	Center for Communications Industry Development
CCP	Chinese Communist Party
CCRI	China Coal Research Institute
CCGT	combined cycle gas turbine
CCT	clean coal technology
CDMA	Code-Division Multiple Access
CEPRI	China Electric Power Research Institute
CIAE	China Institute of Atomic Energy
COSTIND	Commission of S&T and Industry for National Defence
CSIA	China Semiconductor Industry Association
CSIP	China Software and Integrated Circuit Promotion Center
CUSBEA	China–U.S. Biochemistry Examination and Application
CUSPEA	China–U.S. Physics Examination and Application
DFEM	Dongfang Electrical Machine Company
DNA	deoxyribonucleic acid
DVD	digital video disc
EI	Engineering Information
FDA	Food and Drug Administration (U.S.)
GERD	general expenditure on research and development
GIEC	Guangzhou Institute of Energy Conversion
GMO	genetically modified organism
GSM	Global System for Mobile Communications
HGP	Human Genome Project
HPEC	Harbin Power Equipment Company

IC	integrated circuit
ICST	Institute of Computer Science and Technology
ICT	Institute of Computing Technology
IGCC	integrated gasification combined cycle
IPR	intellectual property rights
ISI	Institute for Scientific Information
ISO	International Standards Organization
ISTIC	Institute of Scientific and Technical Information of China
ISTP	Index to Scientific and Technical Proceedings
ITER	International Thermonuclear Experimental Reactor
LME	large- and medium-sized enterprises
MEMS	micro electro mechanical systems
MII	Ministry of Information Industry
MOA	Ministry of Agriculture
MOE	Ministry of Education
MOF	Ministry of Finance
MOH	Ministry of Health
MOST	Ministry of Science and Technology
NCNST	National Center for Nanoscience and Technology
NDRC	National Development and Reform Commission (before 2003 State Development and Planning Commission)
NEMS	nano electro mechanical systems
NERC	National Engineering Research Center
NERCN	National Engineering Research Center for Nanotechnology
NIH	National Institutes of Health (U.S.)
NNSF	National Natural Science Foundation (sometimes also abbreviated as NSFC)
PFBC	pressurized fluidized bed combustion
PRC	People's Republic of China
RIIT	Research Institute of Information Technology
SFDA	State Food and Drug Administration
SGCC	State Grid Corporation of China
SIBS	Shanghai Institutes for Biological Sciences
SOE	state-owned enterprise
SPM	scanning probe microscopy
SWNT	single-walled carbon nanotube

TD-SCDMA	Time Division Synchronous Code Division Multiple Access
TSMC	Taiwan Semiconductor Manufacturing Corporation
USC	ultra-supercritical
USTC	University of Science and Technology of China
WAPI	Wireless Application Protocol Interface
WCDMA	Wideband Code Division Multiple Access
WIPO	World Intellectual Property Office

Introduction

Linda Jakobson

When Director Antti Hautamäki of Sitra, the Finnish Innovation Fund, requested that I assemble a group of specialists to write a "small book" about innovation and high-tech research in the People's Republic of China (PRC), I knew I was wading into deep water. There is nothing small about China. The country is large. The population is gigantic. The national network of science and technology (S&T) research has 5400 governmental institutions, 3400 university-affiliated research institutions, 13,000 research institutions operated by major state enterprises, and 41,000 nongovernmental research-oriented enterprises.[1] The changes over the past 25 years have been enormous. Not only in economic terms—annual income per capita of USD 200 has risen to USD 1700—but also in terms of freedom of choice. A mere 20 years ago, graduating engineers were sent to whichever work unit the state deemed needed their services. They had no alternative other than to obey. Today, state-run universities, research institutes and enterprises compete for the most talented graduates with private firms as well as foreign companies, institutions, and organizations. Graduates might opt to go abroad or start a private company. The decision is their own.

This tremendous transformation of society in such a short time has also brought about mammoth problems. Income disparities have grown rapidly and China has become one of the most unequal societies on the globe. Corruption is rampant. Environmental degradation is colossal. Modernization is being pursued at a stupendous price.

There is truly nothing small about China. But, a small book about a huge topic it would be. Hautamäki would not budge. The smaller the book, he insisted, the greater the chance that it would appeal to a broad audience curious to learn about the evolving Chinese S&T environment and the innovative high-tech research being done in key areas. (Admittedly, the book is not quite as small as Hautamäki hoped.) Hautamäki had a second wish, that the book provide a "view from China." That too is a tall order. The range of views within China on a whole host of subjects has broadened considerably due to the increase in personal freedom and loosening of government controls. This book provides some of these views, drawing on the expertise and opinions of the authors—an eclectic group of specialists.

In many respects, the authors represent a microcosm of the diverse research landscape in China today. Among us, we have three citizens of PRC, a U.S. citizen born in Liaoning province, a U.S. citizen born in the state of Wisconsin, and a European (a Finn born in London). Looking at it from another angle, three of the authors are internationally respected scientists in their fields, one of whom is also a high-ranking S&T policy-maker in PRC; another author is a policy-maker in the field of energy under the influential National Development and Reform Commission; and two of the authors are foreigners, who have been students of China for more than two decades and have lived in PRC for many years. We have all pursued studies abroad. As is evident in the following chapters from the diversity of expressions, styles, approaches, and views, this is truly a Sino-Western joint venture, as was Hautamäki's intention.

Basic terms need to be defined in a book focusing on "innovation" and "high-tech research." Innovation, in the pages that follow, is seen as a process of one idea building upon another, resulting in a new novel product, process or service, which is commercialized and put to practical use. High-tech, according to Wikipedia, is defined as "high technology, technology that is at the cutting-edge and the most advanced currently available."[i] Innovation does not necessarily demand high-tech research, which is important to note bearing in mind the focus of this book on *innovative high-tech research*. In the field of power generation technology, for example, Jiang Kejun writes that Chinese enterprises' innovation capabilities are still weak. He is referring to the capabilities of Chinese enterprises to produce novel, groundbreaking high technology; they could well be highly innovative in applications of existing technology. The line between innovation and innovative high-tech research is easily blurred, and at times the authors do stray into the realm of innovation more generally.

The first chapter strives to paint a picture of China's S&T landscape, assessing the problems China faces as it pursues its goal to become a technological superpower. Two underlying factors are worth keeping

[i] What was perceived as high-tech twenty years ago is not necessarily high-tech today. For definition see en.wikipedia.org/wiki/High_tech.

in mind. On the one hand, the Chinese government is grappling with innumerable challenges simultaneously—the most pressing ones are poverty and health care. Up to five hundred million Chinese exist on under USD 2 a day. Improving the S&T environment is just one of multiple tasks. On the other hand, because the population is aging rapidly, there is a sense of urgency among Beijing's policy-makers to focus on raising the country's technological level. In 2020, one-sixth of the population, an estimated 234 million people, will be over the age of 60. The main goal of the economic reforms launched in the late 1970s was for China to catch up to the West—economically and militarily. The coming two decades are seen as a crucial window of opportunity to make the decisive push before society's resources will be drained caring for the elderly.

The next four chapters provide an overview of high-tech research in key areas and evaluate anticipated progress during the next decade or so. The authors were asked to contemplate a question that intrigues both Chinese and foreigners, anyone wanting to understand where China's economic growth will lead to—technologically, politically and militarily: Is innovative high-tech research expected to emerge from China in the coming ten years?

The focus fields of the book—IT, nanotechnology, energy-related technology, and biotechnology—were chosen based on the relevance they have for China's future S&T development as well as the emphasis that the Chinese government places on them. A common theme that is evident in the chapters is Chinese researchers' awareness of the government's resolve to use the country's limited S&T resources in specifically targeted areas. These sectors receive massive funding through the government's research programs and are the ones in which S&T advances could help solve pressing needs of the population as well as fulfill the aspirations of the government. Chinese scientists are not supposed to pursue science for the sake of science. Scientists should be useful.

However, at the same time, it is important to note that Westerners tend to sometimes mistakenly think that the authoritarian nature of the one-party system in China automatically means that government officials can decree and control everything. They cannot. Deciphering the intricate layers of the permissible and not permissible, as well as the possible and impossible is perhaps one of the biggest challenges for outsiders. This vast country is governed by a single political party, the Chinese Communist Party (CCP), and the ruling elite very much insists that their right to power cannot be challenged. But because of the changes in society and transformed relations between the Party and citizens, for example, the fact that the Party competes for the best and the brightest with many other employers, the government can only dictate up to a certain point. Of course, as elsewhere in the world, by directing funding to specific subareas of chosen technology, the government is without a doubt imposing its will.

Another consideration in determining the focus areas was that IT, nanotechnology, and biotechnology are three of the fields in which there are high expectations that cutting-edge high-tech research will emerge in China. A message that comes across from the authors of the related chapters—in some cases, subtly, in others directly—is the need for a more somber appraisal of the overall state of the innovation system as a whole. Hype abounds about the potential of China's scientists, especially in the fields of IT, biotechnology, and nanotechnology. Despite significant advancement, numerous problems remain. Tackling these issues will take time, as each author explains.

Technology related to the energy sector was chosen even though indigenous, innovative high technology is not expected at any time in the near future to lessen China's heavy reliance on technology transfer from abroad. But even the remote possibility that China would succeed in a breakthrough in developing energy-related high-tech cannot be overlooked. First, China's energy security is a global political issue. Second, China is set to become the largest emitter of carbon dioxide in a few years. The impact of China's growing energy needs on the environment, as well as the effects this increase in energy demand will have on global climate change, are immense. Thus, understanding the technological landscape dominating the energy sector is important to outsiders. Innovative research concepts that are born in China today will be of benefit to others tomorrow. Any research that leads, for example, to substantially bringing down the cost of using existing clean coal technologies is of significance because over two-thirds of China's primary energy use derives from coal. Most of China's small coal-fired plants operate with old-fashioned technology detrimental to the environment. One American energy sector specialist, with years of experience in China, opined in a research interview that it is possible that on the basis of sheer volume, Chinese engineers could surprise the world with innovative research in clean coal technologies. Thousands of Chinese researchers beaver away in this field.

Needless to say, there are other high-tech areas besides the ones focused on in this book, in which Chinese are conducting world-class research. The Chinese government has invested massively in space technology, for example, and has made public its intention to send a Chinese space mission to the moon in the next decade. In its 2006 White Paper on National Defense, Beijing announced similarly high ambitions in the field of dual-use as well as military technologies—these too are beyond the scope of this book. The chapters all concentrate on the civilian applications of high technology.

The book's title is inspired by China's President Hu Jintao,[ii] who in January 2006 proclaimed that China will embark on a new path of "innovation with Chinese characteristics."[2] This is the most fitting translation of China's most recent S&T catchphrase *zizhu chuangxin*,[iii] often translated for lack of a better term as "independent innovation," which highlights the government's ardent desire to strengthen the innovation capabilities of Chinese enterprises. Throughout history, China has developed its own homegrown version of foreign thoughts and terminology, from the transformation of Buddhism that came to China from India in the first century AD to the embracement, in the 1980s, of a "market economy with Chinese characteristics." Also in the realm of science and technology, China is seeking its own unique way to fulfill its goals.

Notes

1. J.S. Xie, W. Blanpied, and M. Pecht, "China's Science and Technology in Electronics, Microelectronics, and Nano-Technologies," in Pecht and Y.C. Chan (eds.), *China's Electronics Industry* (College Park, MD: CALCE EPSC Press, 2005), p. 15.
2. "Hu outlines tasks for building innovation-oriented society," Chinese government's website, 9 January 2006, http://english.gov.cn/2006-01/09/content_151696.htm.

[ii] A word about the order of Chinese names. They are written in the text according to Chinese custom as surname followed by given name. However, at the publisher's request, in the Contents, Notes on Contributing Authors, Foreword and chapter title pages, the names of all contributing authors have been written in Western fashion, i.e. given name followed by surname.

[iii] *Zizhu chuangxin* (自主创新) is ambiguous and not easy to translate precisely. It has been varyingly translated as "independent," "indigenous," or "home-grown" innovation. Co-author Wang Chen suggested "self-motivated innovation," which perhaps best reflects the gist of the term.

1

China Aims High in Science and Technology

An Overview of the Challenges Ahead

Linda Jakobson

1.1 Introduction

China's science and technology prowess is expanding. Whether one examines the number of science and engineering papers that Chinese researchers publish in international journals, the amount of investment made in research and development (R&D) in China, or the number of patents that Chinese are filing, statistics indicate that the science and technology (S&T) capabilities of the People's Republic of China (PRC) are developing rapidly (see Figure 1.1). These advancements are in line with China's leaders' clearly stated goal to make China "an innovation-oriented country" by 2020 and a "world's leading science power" by 2050.[1]

The motives behind China's pursuit of techno-superpower status are manifold. First and foremost, China's political leadership views raising the S&T capabilities of Chinese companies as imperative for economic development to continue. A "shift in the country's economic mode"[2] and significant technological progress are necessary for China to meet the target of quadrupling its 2000 gross domestic product by 2020.[3] As the world economy becomes more knowledge-driven, so will the importance of having the capacity to acquire knowledge as well as to disseminate and apply knowledge. Though China has partially succeeded in moving from labor-intensive, low technology production to high technology manufacturing, it has done so by heavily relying on imported technology. Means of production

Figure 1.1 Selected S&T indicators of China.

	1997	2001	2005
GERD (USD billion)	6.1	12.5	30.1
GERD/GDP (%)	0.64	0.95	1.34
Government S&T appropriation (USD billion)	4.9	8.4	16.4
Government S&T appropriation/total government expenditure (%)	4.4	3.7	3.9
Scientists and engineers (1000 FTEs)	588.7	742.7	1119
Number of patent applications (in China)	114.208	203.573	476.264
S&T papers catalogued by SCI, ISTP, and EI	35.311	64.526	153.374
Exports of high-tech products (USD billion)	16.3	46.5	218.3
Percentage of high-tech products in total exports	8.9	17.5	28.6

Notes: GERD = General Expenditure on R&D, SCI = Science Citation Index, ISTP = Index to Scientific and Technical Proceedings, EI = Engineering Information, FTE = Full Time Employment.

The only exception in the otherwise steady growth in all sectors is the immense growth in exports of high-tech products. The reason for this is at least partially the heavy involvement of foreign and Taiwanese companies in China's high-tech exports.

Source: MOST, China Science and Technology Statistics Databook, http://www.sts.org.cn/sjkl/kjtjdt/index.htm.

for high technology products are to a large extent dependent on the transfer of foreign technology to China. In 2005, 88 percent of China's high-tech exports were produced by foreign (or Taiwanese) owned firms.[4]

So, globally, China is still a borrower, not a creator of new technology. To prosper in an age of knowledge-based economy, a country needs to build a national innovation system within which new knowledge can be transformed into economic growth and welfare.[5] To quote Hu Jintao, China's President and Chinese Communist Party General Secretary: "Real core technologies cannot be purchased but can only be achieved by innovation."[6]

Intertwined with the economic imperatives to upgrade China's S&T capabilities are China's political and military ambitions. These are directly linked to the comprehensive nature of China's aspirations to attain international respect as a major power. While Japan's ascent to economic power in the 1960s and 1970s was distinguished by a low political and military profile, China's emergence as a great trading power has taken place hand in hand with its strengthening political and military might.[i] Science and technology are needed not only to maintain economic growth, but also to boost the nation's international prestige (e.g., by sending a man into space) and modernize the People's Liberation Army. Nationalism is an important component when assessing policies intended to legitimize the Chinese Communist Party's right to govern; hence, advances in science and technology also need to be viewed as a means to enhance national pride. According to Hu Jintao, "Science and technology, especially strategic high technology, is increasingly becoming . . . the focus of competition in comprehensive national strength."[7]

In sum, China aims high in science and technology. China's leaders have made "indigenous innovation" a cornerstone of the country's future development. Several indicators, to be discussed later, show that China has already made impressive strides in S&T in a mere two decades. On the other hand, China's research environment has been criticized as detrimental to individual creativity, corrupt and too politicized;[ii] the quality of Chinese research papers has been deemed low; and S&T policy-makers have been described as overbearing.[8] As always, contradictions abound when analyzing China and where it is heading. How realistic is the ambition to make China an innovative nation by 2020? The chapter addresses this question, providing an overview of the major factors that impact China's pursuit of innovative high technology research.

As is fitting for any assessment of developments in China today, for every argument that indicates that China can achieve its goal, there is a counterargument that says that China will take much longer to

[i] Obviously there are noteworthy differences between China and Japan, which contribute to the comprehensive nature of China's rise: China is home to over a fifth of mankind; China has nuclear capability and in terms of manpower the world's largest armed forces; China is a permanent member of the United Nations Security Council.

[ii] Off-the-record research interviews, upon which this chapter in part draws, were conducted by the author with 103 researchers, officials, and entrepreneurs working in China in the field of S&T in Beijing, Hangzhou, Helsinki, Hong Kong, Shanghai, San Francisco, and Washington DC, 1 November 2005–15 January 2007.

be regarded as an innovative society. China is experiencing complex transition processes that are taking place simultaneously, making any attempt to form a general picture, at best, difficult.

1.2 Looking to the future: National objectives

In February 2006, the Chinese government articulated its S&T goals and strategy for the next 15 years.[9] Dissecting the mammoth "Medium- and Long-term S&T Development Plan" is a mind-boggling task. Wading through 68 prioritized subjects in 11 key fields, 16 mega engineering projects, eight S&T frontiers for research, four major research projects, and eight measures to guarantee the building of an innovative China certainly drives home the message that a wide variety of competing interests and opinions were considered in the plan's drafting process (see Figure 1.2). Some 2000 opinions were initially solicited.

The tangible goals of the 15-year S&T plan can be summarized as follows: China must increase its "indigenous innovation" capacity in order to reduce its reliance on foreign technology to 30 percent or below (from its present reliance of 60 percent);[10] China should be among the top five countries in the world in terms of the number of patents filed for "indigenous" inventions and the frequency of citations in international science papers;[11] China should build several world-class research institutions and universities.

Figure 1.2 Key areas and frontier technology focuses of China's "Medium- and Long-term S&T Development Plan."

Key areas (11)	Frontier technology (8)
Energy	Biotechnology
Water and mineral resources	Information
Environment	New materials
Agriculture	Advanced manufacturing
Manufacturing	Advanced energy
Transportation	Ocean
IT industry and modern services	Laser
Population and health	Aerospace and aeronautics
Urbanization and urban development	
National defense	
Public securities	

What the government views as critical in resolving the country's bottlenecks is perhaps best reflected in the development goals: China should master core technologies in the equipment manufacturing and information industries; catch up with the advanced nations in agriculture-related S&T capabilities; develop energy, energy conservation, and environmental technology; improve the prevention of major diseases; develop modern weaponry; as well as achieve international levels in cutting-edge technologies in information, biology, materials, and aerospace.[12] The need for "indigenous innovation" (*zizhu chuangxin*[iii]) and "leapfrogging" as well as more efficient utilization of China's resources are underlying themes in the plan.

As to the strategy to attain these goals, the government decreed that enterprises must be the driving force behind innovation, and both intellectual property rights (IPR) and standards should be developed as tools to strengthen the competitiveness of Chinese companies. Companies that invest in R&D can expect to receive tax incentives and low-interest bank loans as well as the right to depreciate fixed assets such as facilities.[13] The government has already promised preferential policies to 103 enterprises that it has deemed "innovative."[14] In addition, according to the 15-year S&T plan, government agencies in accordance with government procurement regulations should be obligated to purchase products of "indigenous innovation" that have been developed by domestic enterprises.[15] It is not clear, however, whether these proposed regulations will result in trade-related friction.[16]

According to the Chinese government's plans, expenditure in R&D is to increase substantially. By 2010, investment in R&D will account for 2 percent of GDP, compared to 1.34 percent in 2005. By 2020, the figure should be 2.5 percent of GDP. If reached, this investment level would put China on par with several countries of the Organization for Economic Cooperation and Development (OECD). China would also surpass the European Union in level of R&D investment intensity[17] (see Figure 1.3 and Figure 1.4).

[iii] According to co-author Bai Chunli, *zizhu chuangxin* reflects a goal with a threefold dimension: (1) genuinely original innovation, (2) integration of existing technology, a process, described by Bai, as "one in which many technological innovations are integrated, culminating in the production of a new product," (3) re-innovation, in other words assimilation and improvement of imported technology (discussion with Bai, Beijing, 17 October 2006).

Figure 1.3 R&D expenditure in selected countries.

	Finland 2005	Japan 2004	USA 2004	EU-25 2005	UK 2004	China 2005	India 2003–04
Expenditure on R&D as a % of GDP	3.48	3.18	2.68	1.85	1.73	1.34	0,78 (e)

Note: (e) estimate.
Chinese GERD experienced a rapid annual average growth rate of 19.7% in 2001–2005. This can be partially explained by its low starting point, but the difference between the speed of China's growth and that of, for example, United States (1.7%), EU-25 (1.5%) or Japan (2%) during the same time frame is still enormous.

Source: Eurostat 6/2007; Research and Development Statistics, Government of India: http://www.nstmis-dst.org/RnDPDF/Table%20-%202.pdf.

1.3 Legacies of the past

When Chinese Communist Party leaders in late 1978 approved major reforms focusing on the "four modernizations" (agriculture, industry, science and technology, and defense), China's research and development structure was centrally planned, hierarchal and bureaucratic. It had been developed from the Soviet model.[18] Science and technology resources were, to a great extent, dominated by the military, which led to China developing atomic and hydrogen bombs and ballistic missiles. There was hardly any interaction between institutions, a legacy of the past that still plagues the Chinese R&D system.

The reform period ushered in an era of substantial restructuring of the R&D system. Over the past 25 years, hundreds of research institutes and governmental entities have been merged, abolished, or converted into commercial entities. Indeed, the "past" refers to two phases, pre-1978 and post-1978 to the present, that is the nearly three decades since Deng Xiaoping opened China's doors to Western science and technology. The Chinese government's S&T roadmap for the coming years is entrenched in the S&T reforms of these past decades—"a complex story of 20 years of policy development and institutional reform."[19] During that period the commercialization and internationalization of science took place, not only in China where society at large was undergoing major transformation, but also globally.[20] The emphasis placed on technological self-reliance during the latter part of the Mao era was replaced—others would argue complemented—by an overwhelming reliance on foreign technology.[21] Despite shortcomings,

Figure 1.4 Research and development indicators in selected countries, 2005.

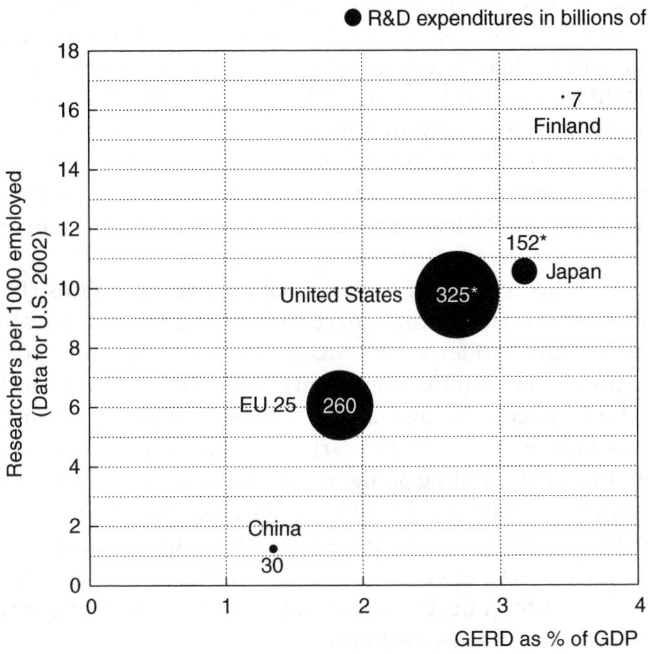

Note: * 2004 data.

Chart shows relative ranking in number of researchers per thousand employed (vertical axis) against gross expenditure on R&D (GERD) as a percentage of GDP (horizontal axis). Absolute expenditure on R&D is also shown (in billions of USD). So, China has relatively few researchers in proportion to the total employed population, while Finland (a country of 5 million inhabitants) has a very high percentage of researchers although China's absolute expenditure on R&D is nearly five times that of Finland. There is an ongoing debate regarding whether one should use real terms or purchase power parity (PPP) when assessing figures related to the Chinese currency. In PPP terms, the figures for China would be significantly higher than in real terms. For simplicity and consistency, all the figures in this book are in real terms.

Source: OECD MSTI 2006-2; Eurostat 6/2007. Figure adapted from OECD Science, Technology, and Industry (STI) Scoreboard 2005.

the S&T reforms of the past two decades have led to the substantial upgrading of China's S&T capabilities, especially in areas such as telecommunications, bio- and nanotechnology.

The legacy of the pre-1978 past includes, on the one hand, the loss of a generation of scholars due to the breakdown of the education system during the Cultural Revolution (1966–76), and, on the other hand, Confucian reverence for education and scholarship, which has propelled Chinese parents to sacrifice for their children's education and spurred Chinese youth into academia. Natural sciences have been and still are held in esteem over social sciences. Moreover, due to the political upheavals of the first 40 years of the PRC, parents have encouraged their offspring to pursue science and engineering (S&E) majors in the hope that scientists and engineers would run a smaller risk than writers or social scientists of being targeted during political campaigns.[iv] Most members of China's ruling Politburo are engineers.

There are other legacies of the past that are relevant to this chapter's focus on the factors that effect China's pursuit of innovative high technology. Besides weak cross-institutional communication, Chinese researchers bemoan the strongly bureaucratic nature of the current S&T system. Though a peer review system for allocating grants has been in place for 20 years, many are critical of the large portion of research funding going to prominent scientists who have good connections with the bureaucracy (privilege based on personal relationships is endemic in Chinese society). They disapprove of basic research being neglected, and point out that even when basic research receives funding, the bureaucrats decree in which prioritized fields basic research should be pursued.

The notion of serving the nation is without doubt one legacy of the past that affects the direction of S&T today. Science and technology officials want to ensure that research is conducted, not for the sake of research but in fields that will benefit the nation. People are no longer expected to do good merely for the sake of the country (or for ideological reasons), as in the past, but are enticed with above average salaries, spacious housing, and overseas travel. Scientists and engineers are now part of the elite, contrary to the Mao era. In research interviews conducted with Chinese scientists, every interviewee said that while he or she, as a Chinese, could accept on a personal level the demand that scientific work should help solve the nation's pressing problems, as a scientist he or she resisted the notion of outside

[iv] Today, university departments teaching economics, finance, and business management compete with engineering and science departments for the brightest students.

interference. Many experienced researchers questioned the practical feasibility of serving the public good while simultaneously fulfilling the government's desire for "indigenous innovation." These tasks require entirely different approaches. The former does not necessarily require cutting-edge research, rather adaptation of technology to local needs and conditions.

The government's call for "indigenous innovation" is also, in some respects, a legacy of the past—yet another catchphrase with a subtle political message. According to a senior academic involved in the final stages of formulating the 15-year S&T plan, focusing on "indigenous innovation" was "a way to highlight the government's disappointment" over China's poor record of domestic technological innovation and overwhelming reliance on foreign technology.[22] As early as 1995, when the government vowed (yet again) to strengthen the nation through science, technology, and education, China's president said, "If we do not have our own autonomous ability to create innovation and just depend on technology imports from abroad, we will always be a backward country. . . . we must remain focused on raising China's ability to do research and development on its own."[23] This aspiration has not materialized. As stated in January 2006 by the *Renmin Ribao* [People's Daily] the mouthpiece of the Chinese Communist Party, "the technological level of China's industries and their capabilities to independently innovate is low."[24]

China's S&T leaders have been adamant in reassuring foreigners that the emphasis on "indigenous innovation" does not mean that China intends to reduce international research cooperation. This seemingly contradictory stance fits in well with the dualistic nature of S&T policy-making (this applies to several fields in China besides S&T, in part due to the transitional nature of the economy and society at large). On the one hand, the government appears stuck in the rut of central planning methodology, and continues to churn out grandiose plans, as it did in the Mao Zedong era. The bulk of central government R&D funding is channeled through five gigantic research programs (Key Technologies, Spark, 863, Torch, and 973) and numerous mega-projects, most of them administered by the Ministry of Science and Technology (MOST), Chinese Academy of Sciences (CAS), or National Natural Science Foundation (NNSF) (see Figure 1.5).[25] On the other hand, the government allows market forces to play a larger role in deciding the fate of state-owned

Figure 1.5 Entities influencing policy-making in field of high-tech research.

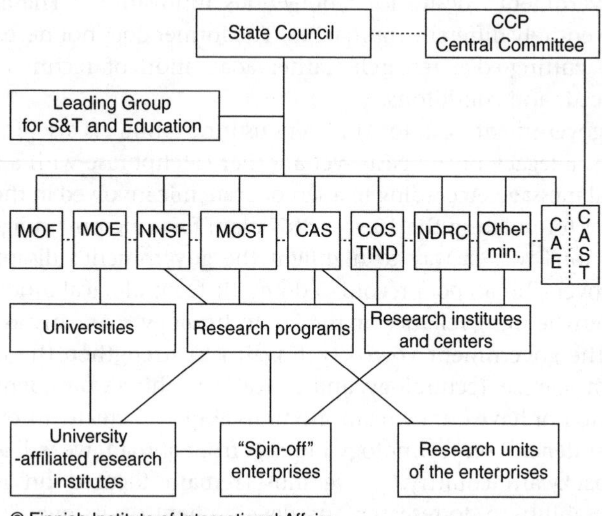

Besides ministries, a whole host of entities affect policy-making, either formally or informally, in the field of high-tech research. For example, CAST (China Association of Science and Technology) is an umbrella organization of S&T academic and industry societies directly under the Central Committee of the Chinese Communist Party (CCP) and though it does not have formal decision-making powers, it is influential (the present Secretary-General is Deng Nan, the late Deng Xiaoping's daughter). Research institutes affiliated with ministries, CAS, universities, or enterprises can also affect policies at the provincial level. There are five major national research programs: The Key Technologies R&D Program was started in 1982 to develop technology urgently needed for industrial upgrading. The Spark program's goal is to introduce advanced technologies to rural areas. "863" (launched in March 1986) is the "High-tech research and development programme" that aims to develop advanced technologies such as IT, biotechnology, and new materials. The Torch program's emphasis is on developing new technology industries. The "National basic R&D program," is "973" started in June 1997. It followed and absorbed the "Climbing program" (initiated in 1989), both of which were intended to make up for the neglect on basic research during the reform era. In February 2007, the Chinese government announced a new funding program for research on key technologies.

enterprises (SOE) and has opened the door to private education. Other contradictory approaches are evident. Chinese S&T officials say that they realize that innovation cannot be dictated from above, while in reality, by stipulating specified research fields and numerically explicit goals, they are attempting to do precisely that. Moreover, when questioned about the feasibility of the ambitious targets set by policy planners, S&T officials tend to answer that plans are just that—plans, and are made to be adjusted.[26] Technology development during the reform period has been characterized as "one of restless change."[27] This resilience, this readiness to adapt to new conditions, and to try alternative paths has to also be seen as a strength of the present Chinese system.

1.4 Creation and diffusion of knowledge

China's S&T landscape is complex, as is befitting a nation as large and diverse as China. The vast majority of high-tech research is concentrated in Beijing and Shanghai. China is living proof of Richard Florida's argument that "the world is spiky."[v] In terms of economic strength and cutting-edge innovation, only the regions around Beijing, Shanghai and Guangzhou are truly consequential. Moreover, innovation strategies in the Pearl River Delta area, encompassing Guangzhou, Shenzhen, Hong Kong, and several cities of over one million inhabitants, focus research on exploring novel applications, rather than basic research.[28]

Though China's S&T landscape is geographically easy to sketch, the picture that emerges of the main actors resembles a complex montage. In addition to the researchers, decision-makers, and educators, important components include foreign direct investment, multinational corporations (MNC), and China-based foreign researchers, most of whom are of Chinese heritage, with academic degrees from foreign universities and work experience abroad, especially from the United States.

1.4.1 The main institutional actors

Knowledge is created and diffused in China by an individual or a group in schools and universities, as well as in research institutes, research departments or research centers. A research unit, regardless of size, is administered by an enterprise (state-enterprise or nonstate enterprise, in other words nongovernmental, private, stockholding, cooperative, joint-venture, or foreign-owned), a university or college, a governmental organization at either the central or local level

[v] Florida argues that contrary to Thomas Friedman's notion that the world is flat, the world of innovation consists of growing peaks and sinking valleys. Innovation is highly concentrated (R. Florida, "The World is Spiky," *The Atlanta Monthly*, October 2005, pp. 48–51).

(e.g., ministry, bureau, or association), a defense-related organization, or a science park. A single instrumental player is the Chinese Academy of Sciences that alone administers some 100 research institutes.

Thousands of high-tech development zones exist in China today: some are no more than a facade for local authorities' real estate speculation or companies' pursuit of tax breaks and export subsidies; some serve their intended purpose and bring together innovative research, entrepreneurship, and venture capital, and are home to successful spin-off companies and small business incubators.[29] Furthermore, beginning in the early 1990s, National Engineering Research Centers (NERC) were established to accelerate the conversion of scientific research in to commercially viable products in several areas, among others electronics and microelectronics, computers, materials, agriculture, energy, and water resources.[30] As of April 2007, the some 140 NERCs were placed under the administration of the National Development and Reform Commission (NDRC).[31]

Science and technology policies are drawn up and overseen by governmental agencies of which the most influential are various leading groups under the State Council, MOST, Ministry of Finance (MOF), NDRC, Ministry of Education (MOE), Ministry of Agriculture, State Commission on Science and Technology for National Defense (COSTIND) and NNSF, the main funding agency of basic research (see Figure 1.5). Most of these entities operate at the central, provincial, and county level. Bureaucratically, on a "ministerial" level, is the prestigious Chinese Academy of Sciences; it is undergoing major reform and has a central role to play in China's technological aspirations.[32] The Chinese Academy of Engineering (CAE) is worth mentioning because many of its 600 academicians are influential policy advisors.

1.4.1.1 The role of enterprises

Enterprises are now proclaimed to have taken center stage of R&D in China. Statistically, R&D spending by enterprises has risen over the past decade. It accounted for 69 percent of China's R&D expenditure in 2005 (see Figure 1.6), which is proportionately on par with R&D spending by enterprises in industrialized countries.[33]

However, in terms of quality, the R&D capabilities of enterprises in China are weak, and it is "unlikely that many Chinese enterprises will develop R&D capabilities in support of novel, science-based technologies in the near future."[34] Leading CAS or key university-affiliated

Figure 1.6 Breakdown of China's R&D expenditure 2004–05.

	2004		2005	
Total expenditure (RMB 100m)	**1996.6**		**2450**	
Institutions	431.7	22%	513.1	21%
LMEs	954.4	48%	1250.3	51%
Universities	200.9	10%	242.3	10%
Others*	409.6	20%	444.3	18%

Note: * Mainly small enterprises.

Total enterprise expenditure (LME + small) constituted nearly 70% of total R&D in China, which is more than for example in the European Union (63%).

Source: OECD, MSTI 2006-2; *China Statistical Yearbook 2006*, pp. 825–827, 830.

research institutes are in a stronger position to conduct innovative high-tech research in prioritized areas, with funding from government programs. Some of these institutes have partnered with high-tech companies. One much-publicized example is Lenovo (formerly Legend), a spin-off company of CAS that in 2004 purchased IBM's personal computer division.

Bearing in mind the focus of this chapter—China's capabilities to pursue innovative high technology research—it is noteworthy that Chinese companies have largely concentrated on product development with little emphasis on hard-core research. Of the 57,786 patents in the invention category that were approved in China in 2006, 43 percent were Chinese (57 percent granted to foreigners), while in the two other categories, utility models and design, 99 percent and 90 percent, respectively, were Chinese patents.[35] Large industrial enterprises spent nearly USD 5 billion in 2004 to purchase foreign technology but only USD 750 million to "absorb and assimilate" technology.[36]

Even the overwhelming majority of Chinese companies who are remarkable business success stories are not high-tech driven. Barring possibly a few solar energy technology companies, none of the top Chinese companies listed on Nasdaq stock exchange, for example, owe their rise to world-class, cutting-edge high technology.[37] Rather, they have mastered a competitive advantage in assembly, design, software, and/or systems engineering. This does not mean that they are merely assemblers, as Chinese companies are often labeled. They are being innovative even though they are not creating new core technology. These Chinese companies too spend on R&D (see Figure 1.7, also p. 54); thus, their input is reflected in data showing the increase of investment by enterprises into R&D.

Except for a few state-owned oil companies and successful non-state IT companies, R&D spending by Chinese companies is low by international standards. Over three-fourths of Chinese large- and medium-sized enterprises (LME) do not even have an R&D department! Total LME spending on R&D in 2005 accounted for 1.54 percent of LME sales revenue.[38] Only five Chinese companies made the list of the world's top 1250 companies in terms of R&D investment.[39]

Figure 1.7 Research and development expenditure flows, 2005.

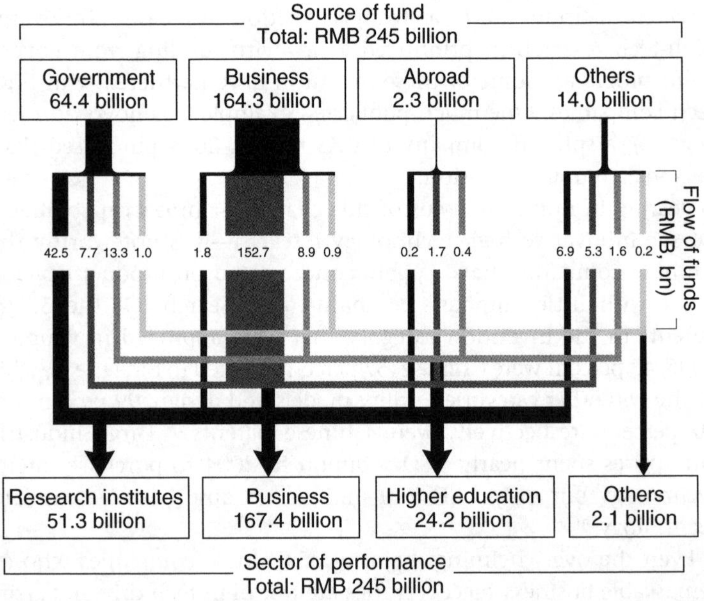

Source of fund
Total: RMB 245 billion
Government 64.4 billion
Business 164.3 billion
Abroad 2.3 billion
Others 14.0 billion
Flow of funds (RMB, bn)
42.5 7.7 13.3 1.0
1.8 152.7 8.9 0.9
0.2 1.7 0.4
6.9 5.3 1.6 0.2
Research institutes 51.3 billion
Business 167.4 billion
Higher education 24.2 billion
Others 2.1 billion
Sector of performance
Total: RMB 245 billion

The bulk of the Chinese government's research expenditure still goes to research institutes, for example, in 2005 research institutes received RMB 42.5 billion from the government, while universities received only RMB 13.3 billion.

Source: MOST, China Science and Technology Statistics Databook 2006, http://www.sts.org.cn/sjkl/kjtjdt/data2006/2006-1.htm; Figure adapted from Schwaag Serger and Breidne, "China's 15-year plan," see endnote 13.

There are multiple reasons for the low R&D investment by Chinese enterprises. When competition is fierce and profits are razor-thin, as is the case in many sectors in China, there is not much left over to invest. The growth of many Chinese companies has been based on cheap labor, a vast domestic market, and imported technology. Even if the technology has often been "second-hand"—not cutting-edge technology—it has been sufficient and has reduced the need for setting up in-house R&D.[40] Purchasing proven imported technology poses less risk than pursuing one's own technology, especially when IPR protection is extremely poor in China. Chinese companies tend to focus entirely on the short-term, reflecting a degree of uncertainty that the Chinese feel about their rapidly transforming society and the possible pitfalls that lie ahead. In addition, among Chinese enterprises and well-off consumers there has been and still is a preference for Western or Japanese technology vis-à-vis home-grown, although attitudes are slowly changing as a result of the success of a few Chinese companies producing world-class brands.

There are other factors that can be attributed to the weak R&D capabilities of Chinese enterprises. Non-state companies, for example, have been given few, if any, meaningful tax incentives to invest in R&D. Nor have they received preferential loans, as have state-enterprises, especially those targeted by the government to be national champions. It remains to be seen how much the innovation capabilities of non-state enterprises will improve toward the end of this decade, in line with the government's commitment in the 15-year S&T plan to provide fiscal support for all companies, regardless of ownership or size.

The main drawback of China's heavy reliance on technology transfer from foreign companies has been the limited spillover of technological know-how to Chinese partners or manufacturers. In the past, R&D centers set up by foreign companies were in private conversations referred to as "PR&D centers" because they were mainly established to create good will with the Chinese authorities and functioned predominantly as product development facilities.[41] However, there have been signs, at least in the pharmaceutical sector, that the tide is turning as multinational companies have started to pursue advanced, even basic research at their R&D centers in China.[42] Related to the question of little spillover from foreign enterprises is the weak record of Chinese efforts and/or capabilities to assimilate foreign technology. This is precisely the reason that the government is so resolute in its promotion of "indigenous innovation."

1.4.2 Human innovators

The basis of an innovative society is undoubtedly its human resources. This is often seen as China's chief asset when assessing its chances of becoming a techno-superpower, along with the steadily increasing investment being made in S&T and higher education. China has the world's second largest workforce of scientists and engineers, after the United States. In 2005, it numbered 2.5 million (vs. about 5 million in the United States and about 8.7 million in EU-25).[43]

1.4.2.1 Educating new talent

China has dramatically expanded university education over the last decade. In 1995, 450,000 Chinese were admitted to university. Ten years later, the number had risen more than fivefold.[44] In 2002, China reached the internationally recognized threshold of mass higher education, "15 per cent of the age cohort," with total enrollment of 16 million students.[45] Science and engineering majors are dominant; 43 percent of all undergraduates study S&E.[46] Noteworthy, however, is that data on S&E bachelor degrees vary greatly depending on the definition of degree. The figure often cited for the number of Chinese S&E graduates per year—600,000—includes at least 200,000 graduates of three-year technical institutes or training programs.[47]

Graduate programs have increased rapidly as well. Since the mid-1990s, graduate enrolments in Master's and PhD programs increased at 30 percent annually. In 2005, nearly 15,000 S&E doctorate degrees were conferred. The new enrolment figure for 2005 was 31,000, which puts China in a position to surpass the global leaders, United States, and Germany, in the number of S&E doctorates annually conferred (approximately 27,000 in the United States, 16,500 in Germany).[48]

Of course, quality matters too. Sheer volume will not make China a technological giant, a fact readily acknowledged by many Chinese education and S&T specialists. An assessment by McKinsey Global Institute evaluated only 10 percent of S&E undergraduates in China to be globally competitive in the outsourcing arena, on the basis of, among other factors, language proficiency and quality of education.[49] The lack of qualified staff is a common complaint among managers of multinational companies and globally oriented Chinese enterprises as well as directors of research institutions. China desperately needs more specialists who are well educated, talented, and able to maneuver within the unique PRC research system.

The quality of education provided in Chinese schools is one of the significant obstacles for China to achieve its goal of building an innovative-oriented society. First, the number of teachers has not increased to correspond with the huge increase in S&E graduates.[50] One professor can be responsible for supervising 75 PhD students. Second, most of the higher education enrolment growth of recent years is a result of burgeoning private education. The quality of education in private colleges varies greatly. Since 1999, public universities have been allowed to establish affiliated, so-called second-tier, for-profit colleges offering degree programs with less stringent entry requirements.[51]

Third, academic corruption in colleges and universities impacts the quality of education. In the last two years, there has been a barrage of articles, reports, and books published in the PRC about corruption linked to postgraduate recruitment, Master's and PhD thesis defense and evaluation, academic title qualification, thesis publishing, and so forth.[52] Furthermore, plagiarism is commonplace. This is, in part, a result of the enormous pressure on everyone, from Master's degree and PhD students to lecturers and professors, to publish academic papers in order to graduate, be promoted, or to receive research grants.

Above all, in spite of several educational reforms since the 1980s, teaching methodology, from primary school all the way to university, is still largely based on rote learning. While students in the PRC excel at tests, they tend not to be initiative takers, critical thinkers, or problem solvers. The Confucian tradition of deferring to authority is not conducive to creativity either. Even at elite universities, graduate students and young lecturers complain that the general atmosphere discourages anyone from questioning professors' views and compels PhD students and junior staff to pursue research according to established patterns with little room for innovative thinking.[53]

Despite the massive infusion of state funding to some forty key universities, it will be years before these institutions fulfill the government's goal of being considered world-class.[54] Most of the best Chinese universities have a few reputable professors and departments that provide high quality teaching and conduct meaningful research.

But two or three excellent departments do not yet make a university world-class. In 2006, not a single Chinese university made the internationally cited Shanghai Jiaotong University's ranking list of the world's hundred top universities.[55] Another widely used, British-compiled global list ranked two Chinese universities in the top hundred.[56]

Finally, the concentration of talent in a small pool drains the energy of the able. Those who are considered competent are given multiple responsibilities. The number of professional hats worn by a Chinese academic is often staggering (and usually evident from his or her business card). One cannot help wondering if any human being can simultaneously be a worthy professor with teaching commitments and the task of supervising several PhD candidates, a competent director of a research institute, an active vice-chairperson of two or three committees advising the government on S&T policies, a participatory member of the board of a university spin-off company, while at the same time keeping abreast of research developments in his or her own field of study and perhaps even publishing or contributing to research. Some or all tasks inevitably suffer. The list of positions tends to be longer if the person has been educated in the West.

1.4.2.2 Bringing back know-how

A cornerstone of Deng Xiaoping's opening policy was to allow Chinese to pursue studies abroad. In the early 1980s outbound students formed a trickle (about 5000 left in 1985) that grew into a steady flow during the 1990s (about 20,000 in 1995), which today resembles a tidal wave. In 2005 alone, nearly 120,000 Chinese went abroad to study.[57] The overwhelming majority of the over one million students who have gone abroad have relied on financial sources other than government funding.[58]

Since 1992, the government has actively tried to entice Chinese who pursued academic degrees and a career abroad to return to work in China. Scores of policies have been introduced to encourage and ease the return of overseas scholars and entrepreneurs. Numerous government programs provide financial support and assistance in finding jobs. Returnees who want to establish their own company are offered tax incentives and other preferential treatment, in some cases considerable support by top officials in securing start-up funding. In research institutes or enterprises they are paid, by Chinese standards, high salaries, and granted spacious housing, a prestigious title, and in some cases large funds to initiate research programs.[59]

Though the number of returnees has increased markedly in the 2000s (over 110,000 returned in 2001–05), hundreds of thousands who left China are still overseas. Many of them are still enrolled in schools abroad, but over 170,000 who no longer study have not returned.[60] This brain drain constitutes a genuine dilemma for the Chinese government. In all but a few fields, the most accomplished specialists have not returned.[61] Rather, returnees in recent years tend to be young BA or MA graduates, who find it difficult to find employment because they lack work experience.[62] The "100 Talents Program" of CAS, one of the most well-funded endeavors to encourage Chinese scholars to the PRC, managed to bring back 778 foreign-based Chinese researchers between 1998 and 2004. But only half had doctorates from foreign universities.[63] In the United States alone, there are 62,500 Chinese-born S&E PhDs who are green card holders;[64] the number doubled in merely four years.[65] This talent is sorely needed in China.

When reputable Chinese scholars do accept positions in China, they do so, nearly without exception, while continuing to hold on to their position abroad. This, in turn, has raised questions of how meaningful their input can be during brief visits to China. Several interviewees asked whether the commitment that a Berkeley or Stanford professor is willing to make for three months each year warrants the high salary and generous perks he is receiving from a Chinese institution.

For Chinese government officials who have publicly encouraged the return of overseas scholars, the question of why so few of the top Chinese researchers abroad have heeded their call is a sensitive one. The reasons for Chinese living overseas to return are manifold and readily discussed: besides the obvious lure of career opportunities in a growing economy and a society buzzing with energy, some make the move for patriotic reasons or because of a desire to care for aged parents or to ensure that their children are fluent in Chinese. There are multiple reasons to stay overseas too, but these are more awkward to expound on. To opt to return to a one-party authoritarian state where no one with certainty knows where the genuine boundary of freedom of speech lies is a leap into the unknown for many who have thrived in an open Western environment. Though the tragic ending of the

Tiananmen democracy movement in June 1989 does not figure in the minds of the under 35-year olds, the Chinese leadership's decision to use force is still mentioned in discussions about the returnee dilemma with middle-aged Chinese. China lost nearly a generation of its best-educated youth in the aftermath of the government crack-down. When, in April 1990, George H.W. Bush issued an executive order allowing Chinese who were in the United States in June 1989 to apply for permanent residency, about 60,000 took him up on the offer.[66] Especially for the Tiananmen generation of academics, the question "what if society does not remain stable?" still lurks in the subconscious.

In spite of the much-publicized efforts of Chinese officials to make the working and living environment suitable for returnees, returning to China after years abroad can prove difficult. On a personal level, life is often not as convenient as the one an academic, previously based in a Western university or research environment, is accustomed to. Returnees tend to feel alienated from their colleagues, partly because of envy stemming from high returnee salaries, but also because of their different outlook on life, based on the years spent abroad. Professionally, returnees within academia share their colleagues' frustration with the hierarchal and bureaucratic way institutions are managed, but are much less patient in expecting changes to take place. Those with working experience in the West find it hard to cope with the ever-present influence of politics in S&T decision-making and the top-down approach with which research institutes are managed. Those who are no longer citizens of the PRC feel slighted when they are barred from research funding (e.g., a major source of funding for S&T researchers, the 863 program, excludes foreign passport holders) and are not eligible to receive various national awards.

Some of the Chinese who have returned from abroad are without question a valuable addition to Chinese science, not only for the knowledge they bring back but also because they have established a network of international connections. However, many returnees have sought employment in multinational companies, which, in turn, is leading to a new kind of "internal" brain drain. Though some of the more adventurous among MNC returnees could one day make the break and start their own enterprises, for the most part, their value for China's S&T aspirations is modest.

While the status of overseas Chinese is complicated, their input is pivotal. Today, thousands of scientists with a Western academic degree and Western academic affiliation work alongside scientists with a solely domestic academic background. Some are citizens of the PRC; some are Taiwanese holding passports of the Republic of China (ROC); others were born in the PRC, then left the country for study and/or work and have returned to the PRC as foreign citizens; others are Overseas Chinese with much older roots abroad, that is they were born abroad and are of ethnic Chinese descent, either second-, third- or even fourth-generation Americans, Brits, Australians, Singaporeans, and so on. Culturally, they are all "Chinese." Yet, there are multilayered nuances of tension associated with their role in the PRC.[vi]

Overseas Chinese have constituted a pillar in the PRC's success. As investors and disseminators of Western technological know-how and management practices their role has been crucial. In the realm of S&T development and education, United States citizens of Chinese descent, in particular, have been instrumental. But though the input of Overseas Chinese is welcomed by PRC officials, it is unclear how these "Chinese compatriots." as they are called in official context, fit in to the government's larger scheme of promoting "indigenous innovation" in order to make China a technological world power. If groundbreaking research is done in a laboratory, situated in the PRC, by a team headed by an American citizen, does this qualify as "indigenous" research? And, if that research is awarded a Nobel Prize, will China claim credit?[67] Also, issues of IPR have yet to be addressed. How do the Chinese and foreign institutions that a head scientist is affiliated with share IPR? It is not inconceivable that if political tensions were to rise between China and Western countries, American and European funding institutions may question where and for whose purposes grant money has been used.

An illuminating example of the confusion and possible misunderstanding that can arise from multiple identities is an April 2006 *People's Daily Online* article, headlined "World's smallest generator developed in China." It fails to mention that the research was done at the Georgia Institute of Technology by a prominent American nanoscientist of Chinese heritage. The scientist was born in the PRC.

[vi] The term for "China" in Mandarin—*Zhong guo* or *Zhong hua*—can mean the country China, but also the Chinese civilization or culture.

In addition to his main job in Atlanta he is a part-time director of a leading nanoresearch center in Beijing.[68]

Naturally, similar questions apply in other countries as well. The globalization of research is becoming the norm, alongside the two-way flows of "transnational capital."[69] and the establishment of "transnational communities" and "global production networks."[70] However, in China, the contradictory position of Overseas Chinese is especially relevant because of, on the one hand, the vital role of the large community of foreigners of Chinese heritage working in the PRC and, on the other hand, the PRC government's emphasis on "indigenous innovation."

1.5 The domestic environment

Throughout the reform period, the leaders of the Chinese Communist Party have made efforts to reform the inner workings of the Party to improve its capabilities to govern China. However, many systemic problems still hamper S&T progress.

1.5.1 Linkages

The so-called stovepipe syndrome that has its roots in the centrally planned economy still plagues China's S&T landscape. Each research entity tends to be an isolated work unit, an island on its own. Even research centers of multinational companies have few ties to the research community at large. Collaboration among research institutes or universities has traditionally been weak, and the same applies to ministries, other government organizations, and state-owned enterprises (this problem affects modernization processes within Chinese society more broadly). In part, this is a legacy of the past, as mentioned earlier. In part, it derives from the present, rigid top-down political system as well as from fierce competition for government funding, talented people, and ultimately, power within the system. In numerous plans and policy outlines, the government encourages cooperation, especially between research institutes and enterprises. In reality, weak linkages remain a major structural challenge.

An obvious consequence of weak collaboration between different entities is that China's limited "scientific resources are scattered, repetitive, and not used efficiently."[71] There is tremendous waste within the system (again, this applies more broadly to Chinese society). Lack of communication between research entities is often accompanied by mistrust, which does not promote the free exchange of ideas. Of course, within academia worldwide rivalry related to funding opportunities are commonplace. The problems are accentuated in China because of the unclear way in which funding decisions are made. When decision-making processes are not transparent and decision-makers are not accountable on the basis of publicized rules that can be challenged, cultivation of personal relations with decision-makers is essential. In research interviews probing the S&T landscape in China, a recurring spontaneous observation was the need for Chinese to study effective system management.

1.5.2 Other systemic problems

The bureaucratic control mentality, inherited from the planned economy, is one of the greatest hurdles China has to overcome in order to transform into a world-class innovative society. The fact that the 15-year S&T plan ended up including four science megaprojects and sixteen engineering megaprojects reflects the bureaucrats' dominance (see Figure 1.8). In the drafting stages of the plan, both behind closed doors and in a few cases in public, several leading scientists made known their staunch opposition to continue the legacy of heavy-handed science planning in the spirit of *liangdan yixing* , the nickname of the 1956–67 program that led to China acquiring nuclear weapons and building satellites.[72] But to no avail.

In four-fifths of the research interviews conducted for this chapter, the Ministry of Science and Technology, in particular, was criticized for its inept bureaucrats, for its attempts to control the direction of research, for its lack of transparency in decision-making, as well as for its overspending on infrastructure and "mammoth projects that give bureaucrats a lot of face." When asked about the role of Minister of Science and Technology Xu Guanhua, a specialist in remote sensing, several interviewees answered that one person cannot change the top-down approach of a ministry that has its roots in the old central-planning system.* "The system is stronger than the person, and the person is molded," a mid-level MOST official said. Several interviewed researchers opined that the only remedy would be the abolishment of MOST and the establishment of a new government body to oversee S&T policies, one based on modern concepts of openness and accountability.

* In April 2007 Xu resigned at age 66 and was replaced with 55-year old Wan Gang, President of Tongji University in Shanghai. Wan is a fuel cell specialist who in 1991 received his PhD in mechanical engineering in Germany before pursuing research at Audi Corporation.

Figure 1.8 Mega projects of the "Medium- and Long-term S&T Development Plan."

Mega science projects (4)	Mega engineering projects (13)*
Protein Science	Core electronic components, high-end generic chips, and basic software
Quantum Research	Extra large scale integrated circuit manufacturing and technique
Nanotechnology	New-generation broadband wireless mobile telecommunications
Development and Reproductive Biology	Advanced numeric-controlled machinery and basic manufacturing technology
	Large-scale oil and gas exploration
	Large advanced nuclear reactors
	Water pollution control and treatment
	Genetically modified new-organism variety breeding
	Drug innovation and development
	Control and treatment of AIDS, hepatitis, and other major diseases
	Large aircraft
	High-definition Earth observation systems
	Manned aerospace and Moon exploration

Note: * The S&T plan states that there are 16 mega engineering projects, but then proceeds to give details for only 13. One can surmise that the remaining three deal with national defense.

The detailed action plan to implement the 15-year S&T plan indicates that MOST will no longer be the main government body in charge of S&T development in China. The role of MOST remains important, but it shares responsibility for implementing the 15-year S&T plan with, among others, MOF, NDRC, and MOE.[73] Implementation of the 15-year S&T plan requires close cooperation between ministries and between central and local government entities—that is a tall order in today's China.

Another major stumbling block is law enforcement, that too a symptom of the systemic problems that the Chinese leadership is grappling with as it attempts to prepare China for the challenges of the twenty-first century. Lack of law enforcement in the realm of S&T translates explicitly into weak enforcement of IPR. Pirating of intellectual property is rampant. Many experts see this as the paramount problem that will deter China from achieving it goals.[74] China will not be looked upon as a favorable location to conduct groundbreaking research before researchers, both Chinese and foreign, can trust the system to protect IPR.

Academic misconduct, including corruption and plagiarism, are systemic problems too. A revealing book written by Professor Liu Ming of Zhejiang University about academic corruption in China documents in detail the customary practices and sums involved in bribing Master's degree and PhD thesis committee members, purchasing academic titles as well as the buying and selling of academic papers. According to Liu, academic corruption is a "typical 'fish rot' syndrome that not only poisons the social environment, but also decays the souls of [future] generations."[75] Liu traces the problems of Chinese science and academia to China's political system, specifically political interference at all levels and the lack of independent organizations. He argues that political interference prevents the peer review system from functioning properly. Lack of reliable qualitative evaluations pushes universities and Chinese science to rely too much on quantitative indicators.

1.6 The international environment

Foreign investment, foreign technology, and foreign expertise have constituted strategic gears of China's modernization drive for over 25 years. They will continue to do so. International research cooperation has been instrumental for China's S&T development. There is every reason to predict that this too will play a significant role in the future. China will further expand its transnational knowledge networks, including tens of thousands of formal and informal types of research cooperation between Chinese and foreign governments, universities, research institutes, and companies, as well as individual researchers.

However, as the global economy moves toward a knowledge-based economy in which the role of technological innovation is central, the economic strength of a nation will increasingly be determined by who owns the technology. Those who control IPR and set technology standards of new products increase their competitiveness. Hence the need for Chinese companies to improve their own innovation capabilities and avoid paying license and royalty fees to foreigners.[76]

The Beijing government aspires to China developing its own standards, at least in prioritized areas. China's national standards strategy advocates that inventions resulting from "indigenous innovation" translate into Chinese standards incorporating Chinese intellectual property. On the other hand, it is not clear whether Chinese enterprises share this goal. They, after all, must adhere to international standards to be competitive in the global market.[77]

China has an additional incentive to strengthen its R&D capabilities: It is restricted from acquiring some so-called dual-use technology because of export controls imposed by the United States and the European Union. Most of these controls were put in place following the Chinese government's suppression of the Tiananmen movement in 1989. The restrictions further strengthen the view among some Chinese policy-makers that outsiders, in particular the United States, are intent on containing China's rise. An example worth mentioning in this context is the consternation among Chinese policy-makers over the U.S. decision in March 2006 to sign a nuclear cooperation agreement with India. Chinese officials interpreted the move as directed against China, which reinforced their view that China must find a way to rely on its own "indigenous" high technology.[78]

In sum, Chinese decision-makers look upon the international environment as both furthering and hindering China's S&T ambitions. As has been the case since the economic reform and opening policies were embraced, the pull-and-push between internationalist and nationalist views among leading Chinese policy-makers is bound to continue. Hence, the balancing of techno-nationalist and techno-globalist objectives will remain a dominant feature of S&T policies in China. China will pursue multiple paths to achieve its goals.

1.7 Conclusion

China has several of the drivers in place to fulfill its technological ambitions. The political elite and scientific community are committed, the government is willing to provide substantial funding to S&T, and the general public approves of placing a high priority on S&T.[79] The speed with which China has acquired the ability to build high quality research and knowledge centers has been phenomenal. A handful of Chinese companies such as Hai'er, Huawei, and Suntech (Shangde) have become world leaders in their own sectors.

However, for a strong national innovation system to emerge in China, four main obstacles need to be overcome. These are raising the quality of Chinese education; reducing the bureaucratic control of S&T policy-makers to facilitate an environment conducive to creativity and cooperation; improving law enforcement to protect IPR; and implementing a system of checks-and-balances to introduce accountability for grant applications, job promotions, thesis approvals, and academic articles. Needless to say, there are numerous other problems China needs to tackle. Many of them are being addressed with targeted policy measures. Chinese leaders are acutely aware of the problems. The inefficient and inappropriate management of research funds is a recurring theme in governmental reports and speeches.[80]

The underlying question is whether a state in which freedom of speech is restricted and stability is placed at the very top of the government's priority list can transform itself into an innovation-oriented society. Can creativity and innovation flourish in an environment that ardently deters people from rocking the boat? That constantly reminds people that the ultimate goal is a harmonious society? That is known to hammer down the nail that sticks out?[vii] The goal of building a harmonious society is a repackaging of the demand for stability.[81] How does one encourage students to challenge authority in the science classroom and prod researchers to disregard recognized models, while simultaneously demanding that they do not question the rules that govern society? These are complex questions to which there are no definite answers. The continuous success of PRC-born students within academia in the United States and Europe is proof of the capabilities of Chinese researchers when they work in an academic atmosphere in which diversity and critical thinking are cherished.

The efforts that China has made and continues to make toward technological excellence will certainly bear fruit. Within the next decade, in some cases perhaps sooner, there will be news of groundbreaking innovative research done in China. There are several pockets

[vii] There are several corresponding Chinese sayings, for example, "The taller the tree, the more wind it attracts;" "A person fears fame, a pig fears becoming stout."

of excellence within such fields as information technology, biotechnology, nanotechnology, and perhaps some other fields of science. But, considering the enormity and extent of the major hurdles China has to overcome to build a comprehensive national innovation system, it is highly unlikely that the transformation could take place in 15 years.

China will proceed to develop its capabilities in the realm of S&T in the same manner it has traversed the reform road since 1978, bit by bit, one step at a time, striving to instill in future generations of Chinese scientists what Premier Wen Jiabao has referred to as an "innovative spirit."[82] Changes will be implemented incrementally. Progress will be piecemeal. At the end of the day, whether or not China achieves its ambitious goals in the realm of S&T depends on how societal reform in China progresses. The S&T landscape faces the same problems as society at large.

Notes

1. "China outlines strategic tasks for building innovation-oriented society," *People's Daily Online*, 9 January 2006, http://english.peopledaily. com.cn/ 200601/09/eng20060109_233967.html. "China strives to be science power," Chinese government's website, 9 January 2006, http://english.gov.cn/ 2006-02/09/content_184335.htm.
2. "Chinese President stresses urgency of 'shifting economic growth mode,'" *People's Daily Online*, 17 January 2006, http://english.peopledaily. com.cn/200601/17/eng20060117_235924.html.
3. The target of quadrupling the 2000 level GDP by 2020 was articulated at the Sixteenth Party Congress of the CPC in November 2002. "China to Quadruple GDP in 2020 from 2000," *People's Daily Online*, 8 November 2002, http://english.peopledaily.com.cn/200211/08/eng20021108_106504. shtml.
4. Ministry of Commerce, "海关公布年度出口 200 强 华为居民营企业首位" [Huawei no. 1 in annual Customs-published list of 200 strongest exporters], 16 June 2006, http://foreigntrade.mofcom.gov.cn/aarticle/c/ 200605/20060502198530.html.
5. S. Schwaag Serger and E. Widman, "Competition from China-opportunities and challenges for Sweden," Östersund: Swedish Institute for Growth Policy Studies (ITPS), 2005, p. 13, http://www.itps.se/Archive%2FDocuments% 2FEnglish%2FPublikationer%2FRapporter%2FAllm%E4nna%2FA2005%2F A2005_019_eng.pdf.
6. "China outlines strategic tasks for building innovation-oriented society," *People's Daily Online*, 9 January 2006, http://english.peopledaily.com.cn/ 200601/09/eng20060109_233967.html.

7. "President Hu calls on scientists to help achieve goal of innovation nation," *People's Daily Online*, 6 June 2006, http://english.peopledaily.com.cn/200606/06/eng20060606_271360.html.
8. Author's research interviews during 2006. See also M.M. Poo, "大科学和小科学" [Big Science, Small Science], *Nature*, 432, China Supplement (18 November 2004), pp. 18–23; Y. Rao, B. Lu, and C.L. Tsou, "中国科技需要的根本转变：从传统人治到竞争优胜体制" [A Fundamental Transition from Rule-By-Man to Rule-By-Merit: What Will be the Legacy of the Medium- and Long-term S&T Development Plan?], *Nature*, 432, China Supplement (18 November 2004), pp. 12–17; Poo, "Cultural reflections," *Nature*, 428, English-language China supplement, (11 March 2006), pp. 204–205; R. Wu, "Making an Impact," *Nature*, 428, English-language China supplement, (11 March 2006), pp. 206–207. The November 2004 *Nature* supplement in Chinese was initially quoted in the PRC media until its distribution was banned in China, and Chinese editors were allegedly told not to refer to it nor comment on the draft of the 15-year S&T plan (X. Hao and Y.D. Gong, "China Bets Big on Big Science," *Science*, 311 (17 March 2006), p. 1549).
9. State Council of the People's Republic of China, "国家中长期科学和技术发展规划纲要" [Outline of National Medium- and Long-term S&T Development Plan (2006–2020)], 9 February 2006, http://www.gov.cn/jrzg/200602/09/content_183787.htm. The English translation of the plan is a 23,000-word document, see for example Xinhua (New China News Agency), "Text of PRC Medium- and Long-term S&T Program Guidelines for 2006–2010," 9 February 2006 (English translation by Open Source Center).
10. According to MOST, reliance on foreign technology is calculated using the following index: reliance on foreign technology = expenditure on importing foreign technology / (domestic R&D expenditure + expenditure on utilizing domestic technology + expenditure on importing foreign technology). See question posed by Yu Fengliang on 17 August 2006 at link: http://appweblogic.most.gov.cn/gzwd/gzwd_jsjg.jsp?Tid2=002&page=28.
11. In 2006 China was the eighth largest country in Patent Cooperation Treaty filings, see World Intellectual Property Organization (WIPO), "Record Year for International Patent Filings with Significant Growth from Northeast Asia," 8 February 2007, http://www.wipo.int/edocs/prdocs/en/2007/wipo_pr_2007_476.html.
12. Outline of National Medium- and Long-term S&T Development Plan (2006–2020).
13. These advantages were all mentioned in the draft of the revised S&T Progress Law under consideration in early 2007. Ministry of Science and Technology, "法制办就科学技术进步法（修订草案）征求意见" [Legal Office seeks opinions on the revised draft of the S&T Progress Law], 22 March 2007, http://www.most.gov.cn/yw/200703/t20070326_42340.htm.

14. Ministry of Science and Technology, "创新型企业试点工作初见成效" [Preliminary results of the pilot project on innovative enterprises], 26 February 2007, http://www.most.gov.cn/kjbgz/200702/t20070226_41515.htm.
15. S. Schwaag Serger and M. Breidne, "China's 15-year plan for science and technology—a critical assessment," paper presented at conference "New Asian Dynamics in Science, Technology and Innovation," Gilleleje, Denmark, 27–29 September 2006, p. 15.
16. China has not yet signed the Government Procurement Agreement (GPA). Until it joins the GPA, China is technically not obligated to refrain from protecting its domestic industries. Upon joining the GPA, however, it then must not discriminate against foreign suppliers above certain thresholds and within the sectors/entities covered in the agreement, but can continue to do so in those areas not covered. China committed to join the GPA as soon as possible in its WTO Accession Protocols (author's correspondence with Louisa Chiang of the Trade Facilitation Office at the U.S. Embassy in Beijing, 14 February 2007).
17. Eurostat, "In relation to GDP, EU27 R&D expenditure stable at 1.84% in 2005," 6/2007, 12 January 2007. In real terms EU27 spent about 200 billion euro on R&D in 2005, while R&D intensity for EU27 was 1.84 percent of GDP, the same as in 2004. While the Lisbon Summit Strategy goal of 3 percent appears impossible to fulfill by 2010 (or even 2020) for the whole EU, individual countries, for example, Sweden and Finland have already passed this target; in 2005 Sweden used 3.86 percent and Finland 3.48 percent of GDP on R&D.
18. For a summary of the history of the PRC's technology policy, see K. Walsh, "Science, Technology, and High-Tech Development in China," in *Foreign High-Tech R&D in China* (Henry L. Stimson Center, 2003), pp. 35–72.
19. R.P. Suttmeier, C. Cao, and D.F. Simon, "'Knowledge Innovation' and the Chinese Academy of Sciences," *Science*, 312 (7 April 2006), p. 58. I am grateful to Richard P. Suttmeier for reminding me of the importance of viewing S&T developments in China against the backdrop of the steps taken over the past 20 years.
20. For an overview of changes in the S&T landscape in China during the reform period, see P. Suttmeier and C. Cao, "China's Technical Community," in E. Gu and M. Goldman (eds.), *Chinese Intellectuals between State and Market* (London: Routledge Curzon, 2004), pp. 138–157.
21. For a discussion of self-reliance in China's development strategy, see D. Kerr, "Has China Abandoned Self-reliance?" *Review of International Political Economy*, 14: 1 (February 2007), pp. 77–104; R.P. Suttmeier, "A New Technonationalism? China and the Development of Technical Standards," *Communications of the ACM*, 48: 4 (April 2005), pp. 35–37.
22. Research interview with senior Chinese academic who serves on several governmental advisory committees, Beijing, 25 April 2006. For a discussion of the controversy surrounding *zizhu chuangxin* and the Chinese academic community's reactions to the new policy emphasis, see L. Jakobson,

"China's new R&D focus: The meaning and intent of 'indigenous innovation,'" paper presented at conference "New Asian Dynamics in Science, Technology and Innovation," Gilleleje, Denmark, 26 September 2006.

23. Jiang Zemin's speech at "Chinese National Conference on S&T," 26 May 1995. Quoted in http://www.globalsecurity.org/military/library/report/1996/stpol1.htm.
24. "高自主创新能力是推进结构调整的中心环节" [Improving the ability of indigenous innovation is the central link to structural adjustment], *Renmin Ribao* [People's Daily], 18 January 2006.
25. For an overview of China's research programs, see J. Sigurdson, *Technological Superpower China* (Cheltenham: Edward Edgar, 2005), pp. 8–11, 38–53; C. Cao, "China Planning to Become a Technological Superpower," EAI Background Brief no. 244 (Singapore: East Asian Institute, May 2005), pp. 14–15; R.P. Suttmeier and C. Cao, "China's Technical Community," pp. 144–146; Chinese government's website "National Programs for Science and Technology," 9 February 2006, http://english.gov.cn/2006-02/09/content_184156.htm.
26. Author's research interviews with eleven MOST officials, 3 January–24 September 2006.
27. B. Naughton and A. Segal, "China in Search of a Workable Model. Technology Development in the New Millennium," in W.W. Keller (ed.), *Crisis and Innovation in Asian Technology* (West Nyack, NY: Cambridge University Press, 2003), p. 162.
28. Needless to say, there are individual institutions elsewhere in China that conduct first class high-tech research, primarily in Chengdu, Chongqing, Dalian, Jinan, Shenyang, Tianjin, Wuhan, and Xian. For a regional overview of S&T developments in China, see J. Wildson and J. Keeley, "China: The Next Science Superpower?" (London: Demos, 2007), pp. 34–37.
29. In 2004, there were 53 state-level high-tech development zones, housing a total of over 30,000 enterprises. Company size varies greatly. Thirty had annual production values of over RMB 10 billion (USD 1.3 billion), more than 200 of over RMB five billion (USD 650 million), and 3000 of over RMB 100 million (USD 13 million). "New and High-Tech Development Zones," http://www.china.org.cn/english/features/China2005/142157.htm.
30. J.S. Xie, W. Blanpied, and M. Pecht, "China's Science and Technology in Electronics, Microelectronics, and Nano-Technologies," in Pecht and Y.C. Chan (eds.), *China's Electronics Industry* (College Park, MD: CALCE EPSC Press, 2005), p. 17.
31. NDRC, "国家工程研究中心管理办法" [Managing NERCs], 5 March 2007, http://www.86148.com/chinafa/shownews.asp?id=9907; see also 国家工程技术研究中 心 – 中心简介 [Introduction to National Engineering Research Center], http://www.cnerc.gov.cn/aboutus/index.aspx.
32. For an overview of the Knowledge Innovation Program, an ongoing reform process of the Chinese Academy of Sciences, see R.P. Suttmeier, C. Cao, and D.F. Simon, "China's Innovation Challenge and the Remaking

of the Chinese Academy of Sciences," *Innovations* (Summer 2006), pp. 78–97.

33. Ministry of Science & Technology, "RMB 245 Billion R&D Expenditure," Newsletter no. 451, 20 September 2006. http://www.most.gov.cn/eng/newsletters/2006/200610/t20061027_37210.htm; OECD, *Main Science and Technology Indicators*, December 2006.
34. See Suttmeier, Cao, and Simon, "China's Innovation Challenge," p. 80.
35. State Intellectual Property Office, "Grants for Three Kinds of Patents Received from Home and Abroad," 15 January 2007, http://www.sipo.gov.cn/sipo_English/statistics/gnwsqnb/2006/200701/t20070115_126914.htm.
36. "经济观察：九大问题挑战 '创新型国家'" [Economic observations: Nine big questions challenging an "innovative nation"], *Renmin Ribao* [People's Daily], 9 January 2006.
37. Research interview with Professor Martin Kenney who has done extensive research on venture capital markets, San Francisco, 29 March 2006.
38. National Bureau of Statistics of China, *China Statistical Yearbook 2006*, (Beijing: China Statistics Press), p. 827.
39. The five Chinese companies to make the list: PetroChina (185), China Petroleum & Chemical (260), ZTE (298), Semiconductor Manufacturing (701), CNOOC (963). Department of Trade and Industry (U.K.), "The R&D Scoreboard 2006," vol. 2, p. 84, http://www.innovation.gov.uk/rd_scoreboard/. It is worth noting that Huawei spends more on R&D than any Chinese company that made the list; the RDI Scoreboard is based on survey results, so it is possible that Huawei simply did not answer the survey. Moreover, Semiconductor Manufacturing is majority foreign-owned.
40. S.L. Gu and B-Å. Lundvall, "Policy Learning in the Transformation of the Chinese Innovation Systems," in Lundvall, P. Intarakumnerd, and J. Vang (eds.), *Asia's Innovation Systems in Transition* (Cheltenham: Edward Elgar Publishing Ltd, 2006), p. 301.
41. A study headed by the Science Counselor of the Swedish Embassy in Beijing estimated that only 30 large MNCs were conducting innovative research in China. S. Schwaag Serger, "From Shop Floor to Knowledge Factory," in M. Karlsson (ed.), *The Internationalization of Corporate R&D, Leveraging the Changing Geography of Innovation* (Öresund: The Swedish Institute for Growth Policy Studies, 2006), p. 245, http://www.itps.se/Archive/Documents/Swedish/Publikationer/Rapporter/Allm%C3%A4nna/A2006/Kap10_A2006_007.pdf.
42. Copenhagen-based Novo Nordisk, the first multinational pharmaceutical company to open an R&D center in China, was a pioneer and, from its establishment in 2001, assigned its Beijing center tasks including basic research in genetic engineering of bacteria and antibody technology (correspondence with E. Boel, Vice-President, Novo Nordisk, 7 November 2006). In late 2006, Boston-based Novartis announced it would conduct basic scientific research in China, by investing USD 100 million in a new drug discovery facility in Shanghai (G. Dyer, "Novartis to open research facility in China," *Financial Times*, 6 November 2006).

43. The size of the S&E workforce naturally varies based on the criteria used to define scientist or engineer. The figure for China is for "scientists and engineers engaged in R&D activities" in 2005 (*China Statistical Yearbook 2006*, p. 825). The U.S. figure 4.9 million is for "those in S&E occupations with at least a bachelor's degree" in 2003 (NSF *Science and Engineering Indicators 2006*, Table 3–5). The EU-25 figure is from Eurostat, "Measuring gender differences among Europe's knowledge workers," *Statistics in Focus, Science and Technology*, 12/2006, p. 3, http://epp.eurostat.ec.europa.eu/cache/ITY_OFFPUB/KS-NS-06-012/EN/KS-NS-06-012-EN.PDF.
44. *China Statistical Yearbook 1996*, p. 634; *2006*, p. 803. The figure 450,000 is for new admissions to four-year undergraduate programs. In addition, 480,000 new students were admitted to "colleges for professional training" in 1995; 2,680,000 in 2005.
45. R. Hayhoe and Q. Zha, "Becoming World Class: Chinese Universities Facing Globalization and Internationalization," *Harvard China Review* (Spring 2004), p. 88.
46. *China Statistical Yearbook 2006*, p. 805.
47. *China Statistical Yearbook 2005*, p. 697 gives the figure 576,627 for S&E graduates from undergraduate programs in 2004. According to a Duke University study, in 2004 China awarded 376,415 bachelor degrees in engineering, computer science, and information technology, compared to 137,437 in the United States and 139,000 in India (G. Gereffi, V. Wadwha, and B. Rissing, "Framing the engineering outsourcing debate: comparing the quantity and quality of engineering graduates in the United States, India and China," paper prepared for SASE 2006 conference "Constituting Globalisation: Actors, Arenas, and Outcomes," in Trier, Germany, 27 June 2006, pp. 6–8).
48. Data for Germany from 2003. Science and Engineering doctorates conferred in Britain 8810 (2003); Japan 7581 (2003); Finland 922 (2002); in EU-14 (Luxembourg missing) 40,776 (2002). Data for China from *China Statistical Yearbook 2006*, p. 802; 2005, p. 694. All other data from NSF *Science and Engineering Indicators 2006*, Tables 2–40, 2–42.
49. McKinsey Global Institute, "The Emerging Global Labor Market: Part 1—The Demand for Offshore Talent in Services," June 2005, p. 24.
50. See Y. Lin, "The Grim Situation in China with Regard to Building a Teacher Base and Rejuvenating the Higher Education System," *Chinese Education and Society*, 38: 4 (July/August 2005), p. 33; see also Gereffi, Wadhwa, and Rissing, "Framing the Engineering," pp. 14–15.
51. World Education News and Review, "International Rankings and Chinese Higher Education Reform" (New York: World Education Services, October 2006), http://www.wes.org/ewenr/PF/06oct/pfpractical.htm.
52. For a comprehensive overview on academic corruption in China, see chapter 2 ("Criticism of Academic Morality," pp. 13–26), in book by Professor Liu Ming of Zhejiang University: 刘明, 学术评价制度批判 [Criticism of the Academic Evaluation System], 长江文艺出版社 [Changjiang Literature and Art Publishing House], 2006. I am grateful to David Cowig for pointing out this enlightening book. See also "高校的非典型腐败"

[Abnormal corruption of higher education] 中国新闻周刊 [*Newsweek*, Chinese edition], 20 March 2006.

53. Since 1988 to the present, the author has regularly either conducted research interviews or had informal discussions about the quality of teaching and educational reform at five of China's top ten universities: Beijing, Tsinghua, Fudan, Zhejiang, and Renmin University. See also e.g. P. Mooney, "The Long Road Ahead for China's Universities," *Chronicle of Higher Education*, 52: 37 (19 May 2006), pp. 51–52.
54. Nearly 40 universities presently receive prioritized government funding via "Project 985" that aims to hasten their transformation into world-class universities. In May 1998, "985" was founded and supplemented the Chinese government's earlier "211 program," intended to aid 100 universities to achieve world-class status, by providing even greater funding (see World Education News and Review, "International Rankings.") Furthermore, one of the goals of the Knowledge Information Program of CAS is to create 30 internationally recognized research institutes by 2010, with five recognized as world leaders (see Suttmeier, Cao, and Simon, "Knowledge Innovation," p. 58).
55. "Academic Ranking of Top 500 Universities 2006." Because of its methodology emphasising S&T, the Shanghai Jiaotong University list ranks almost exclusively research universities and not liberal-arts colleges: http://ed.sjtu.edu.cn/rank/2006/ARWU2006_Top100.htm.
56. "THES-QS World University Rankings 2006/07," published by the *Times Higher Education Supplement*. Beijing University ranked 14th, Tsinghua 28th: http://www.topuniversities.com/worlduniversityrankings/results/2006/.
57. *China Statistical Yearbook 2006*, p. 801.
58. For an overview of funding of overseas students, see C. Li, "Coming Home to Teach: Status and Mobility of Returnees in China's Higher Education," in C. Li (compiled), *Bridging Minds across the Pacific* (London: Lexington Books, 2005), p. 77. The figure over one million for total students who have left China to study abroad by the end of 2006 is an estimate based on *China Statistical Yearbook 2006*, p. 801 and MOE report "2005年留学工作年鉴" [Overseas Students' Work Report 2005], http://www.moe.gov.cn/edoas/website18/info12181.htm.
59. For an overview of Chinese government policies to reverse the brain drain, see D. Zweig, "Competing for Talent: China's Strategies to Reverse the Brain Drain," *International Labour Review*, 145: 1–2 (2006), pp. 65–78.
60. Over 540,000 students left China for study abroad 2001–05 and could not be expected to have finished their degrees yet (*China Statistical Yearbook 2006*, p. 801). The figure 170,000 is from returnee data up till end of 2003 (MOE report "Work Related to Students and Scholars Studying Abroad," http://www.moe.gov.cn/english/international_2.htm).
61. Several China-based academics as well as Chinese S&T officials put forward this stance in research interviews conducted by the author in China during 2006. See also Zweig, "Competing for Talent," pp. 78–80; Wildson and Keeley, "China: The Next," p. 29.

62. Personal correspondence with David Zweig, 7 January 2007.
63. See Suttmeier, Cao, and Simon, "China's Innovation Challenge," p. 83.
64. The author first became aware of this figure during the power presentation of R.P. Suttmeier and C. Cao at China–U.S. Forum on S&T Policy, Beijing, 16 October 2006. The figure, based on 2003 data, is from *NSF Science and Engineering Indicators 2006*, Table 3–18.
65. Data from 1999. European Commission, "European PhD holders in the U.S.," Third European Report on S&T Indicators 2003, March 2003, ftp://ftp.cordis.europa.eu/pub/indicators/docs/3rd_report_snaps3.pdf.
66. L. Jakobson, *A Million Truths. A Decade in China* (New York: M. Evans, 1998), p. 258. Furthermore, about 11,000 mainland Chinese students in Canada received permanent residency.
67. For a discussion of the "Nobel Prize mania" in China, see C. Cao, "Chinese Science and the 'Noble Prize Complex'," *Minerva*, 42 (2004), pp. 151–172.
68. "World's smallest generator developed in China," *People's Daily Online*, 25 April 2006, http://english.people.com.cn/200604/25/eng20060425_261054.html; Z.L. Wang and J.H. Song, "Piezoelectric Nanogenerators Based on Zinc Oxide Nanowire Arrays," *Science*, 312 (14 April 2006), pp. 242–246.
69. S. Rosen and D. Zweig, "Transnational Capital: Valuing Academic Returnees in Globalizing China," in C. Li (compiled), *Bridging Minds across the Pacific* (London: Lexington Books, 2005), p. 112.
70. A.L. Saxenian, "Transnational Communities and the Evolution of Global Production Networks: The Cases of Taiwan, China, and India," *Industry and Innovation*, 9: 3 (December 2002), pp. 183–202.
71. Y. Zheng and M.J. Chen, *China Plans to Build an Innovative State* (Nottingham: China Policy Institute, University of Nottingham, June 2006), p. 12.
72. Author's research interviews with 12 Chinese who participated in the drafting of the 15-year S&T plan, 1 November 2005–15 January 2007; three of the interviewees were involved in the final stages of the plan's approval process. See also Poo, "大科学和小科学" [Big Science, Small Science], pp. 18–23; Hao and Gong, "China Bets Big on," pp. 1548–1549.
73. "国务院办公厅关于同意制订《实施〈国家中长期科学和技术发展规划纲要〉的若干配套政策》实施细则的复函" [Letter of reply by General Office of the State Council regarding the approved formulation of list of detailed rules and regulations for "Implementing a number of policies for the 'National Medium- and Long-term S&T Development Plan'"], http://www.gov.cn/gongbao/content/2006/content_310755.htm. The first installment of support policies were published four months after the 15-year S&T plan was officially unveiled. The NDRC was assigned responsibility for 29 so-called support policies, the MOF for 21, MOST for 17, and the MOE for 9. For a discussion of the 15-year S&T plan's supporting policies and role of MOST, see Schwaag Serger and Breidne, "China's 15-year plan," pp. 10–14.

74. See for example A. Stevenson-Young and K. DeWoskin, "China Destroys the IP paradigm," *Far Eastern Economic Review*, 168: 3 (March 2005), pp. 9–18.
75. See M. Liu, 学术评价制度批判 [Criticism of the Academic Evaluation System], p. 16.
76. For a detailed discussion of China's national standard strategy, see R.P. Suttmeier, X.K. Yao and A.Z. Tan, *Standards of Power? Technology, Institution, and Politics in the Development of China's National Standards Strategy* (Seattle: National Bureau of Asian Research, June 2006), pp. 1–16, 28–39.
77. Discussions with Chris Lanzit who works in the field of standards as Executive Director of the Beijing office of Consortium of Standards and Conformity Assessment (CSCA) that encompasses four U.S.-based standards development organizations (API, ASME, ASTM International, and CSA America). See also p. 46 and pp. 65–66 in Chapter 2. Suttmeier, Yao, and Tan also allude to competing interests [of enterprises] that do not necessarily coincide with the political will of policy-makers (see Suttmeier et al., *Standards of Power?* p. 37).
78. The historic pact on nuclear cooperation between the United States and India, which commits the United States to share nuclear reactors, fuel, and expertise with India, was announced on 2 March 2006 when the author was conducting research interviews in Shanghai. Both in Shanghai and during discussions with S&T officials during the following weeks in Beijing, the U.S.-India deal was brought up by the interviewees as proof of the U.S. intent to contain China's rise.
79. See R. Silberglitt, P.S. Anton, D.R. Howell, and A. Wong, "The Global Technology Revolution 2020, In-Depth Analysis," National Security Research Division, Rand Corporation, 2006, pp. 78–83.
80. For example, a progress report published by MOST in January 2007 reflected the government's dissatisfaction with the inefficient use of resources allocated to R&D. According to the report, progress in the previous 12 months had been modest despite a substantial increase in government funding (MOST, "徐冠华部长在 2007 年全国科技工作会议上的报告" [Minister Xu Guanghua's Report at the National S&T Work Conference 2007], 29 January 2007, http://www.most.gov.cn/tztg/200702/t20070209_41241.htm). Moreover, during the National People's Congress in March 2007, a lawmaker accused top universities of misusing research funds to build luxury buildings (H.P. Jia, "Top Chinese universities 'mis-spending' funds." SciDev.Net, 8 March 2007, http://www.scidev.net/News/index.cfm?fuseaction=readNews&itemid=3465&language=1).
81. I am grateful to Denis Fred Simon for reminding me of the parallel.
82. See, e.g., Graduate School of CAS, "温家宝强调：要把教育摆在优先发展的战略地位" [Wen Jiabao emphasizes: Education must be given strategic status in order to enhance development], 29 November 2006, http://news.gscas.ac.cn/info_www/news/detailnewsb.asp? infono=8579.

2
China's Push to Innovate in Information Technology

Arthur Kroeber

2.1 Overview

In the last decade, China has made a startling leap into the front ranks of producers of information technology (IT) goods. Between 1996 and 2004 the value of China's exports of electronic and communications equipment, and other goods classified as "high-tech" by Chinese customs, grew more than ninefold, from USD 19 billion to USD 180 billion. In 2004, China overtook the United States as the world's leading exporter of information-technology goods (see Figure 2.1). The structure of China's IT manufacturing is broad, with substantial exports of computers, consumer audio-video equipment, telecommunications equipment, and components.

This astonishing ascent has generated many headlines proclaiming China a "high-tech powerhouse." In reality it is no such thing. The three salient facts about China's IT hardware industry are:

- *Production is mainly last-stage, low-value assembly.* China has become the location of choice for final assembly of IT products whose components are manufactured elsewhere in Asia, and whose core technologies are designed in North America, Europe, and Japan. Most of the value of an IT product assembled in China is captured by these core technology designers. The next biggest chunk is captured by specialized component makers, mainly in Taiwan, South Korea, and Southeast Asia. The Chinese value share is generally estimated at 10–15 percent—and the majority of that value is captured by the Asian subcontractors of multinational firms, not by domestic companies.

Figure 2.1 Global imports and exports of high-tech goods (Figures in billion USD).

Imports	**1996**	**2000**	**2004**
United States	150	238	235
EU-15	106	167	226
Japan	48	67	73
China	**17**	**51**	**149**
Exports	**1996**	**2000**	**2004**
United States	124	182	149
EU-15	73	111	139
Japan	103	124	124
China	**19**	**47**	**180**
Balance	**1996**	**2000**	**2004**
United States	−27	−56	−86
EU-15	−32	−56	−87
Japan	55	57	51
China	**2**	**−4**	**31**

Note: Data for EU exclude intra-EU trade.

Source: "OECD finds that China is biggest exporter of Information Technology Goods in 2004, surpassing US and EU," 12 December 2005, http://www.oecd.org/document/8/0,2340,en_2649_201185_ 35833096_1_1_1_1,00.html.

China is the last station on a great Asian manufacturing conveyor belt and its contribution in value terms is the smallest.[1] Government laboratories, notably at the Chinese Academy of Sciences (CAS), have had some success in building high-speed computers for use in China's aerospace and genomics programs. These efforts are impressive technical achievements but do not represent meaningful innovation.

• *Foreign firms dominate.* Foreign firms accounted for 88 percent of China's high-tech exports in 2005.[2] Unlike Japan and South Korea, China has not produced domestic companies that are anywhere near exercising leadership in global IT markets.

• *Innovation is limited.* The comparative advantage of Chinese IT production is in the commoditization of technology: that is, producing in large volumes and at very low cost. This dovetails with a major objective of China's overall development strategy, which is to make IT widely available at the lowest possible cost. The extremely high rate of technology diffusion means that in the long term, China's prospects for becoming a center of innovation are bright. Yet China's overwhelming comparative advantage in low-cost production, combined with poor protection for intellectual property and a defective system for commercializing the results of basic research, mean that it is unlikely to be a source of significant innovation in the short- to medium-term.

The relatively low value of Chinese IT product assembly, and the dearth of domestic innovation, is a source of intense concern among government policymakers. They have crafted a variety of incentive policies, and directed substantial research investments which are frequently labeled as aiming to promote innovation. In reality, however, these policies have little to do with innovation—the sources of which are poorly understood by Chinese policymakers—and much to do with increasing the market share and profits of domestic (often state-owned) firms.

This chapter will examine the objectives of national IT research in China, along with the government policies and funding mechanisms that support this research, as well as the intellectual property rights (IPR) and standards environment. It then surveys the major sectors of China's IT industry and assesses the prospects for innovation in each.

2.2 National objectives of IT research in China

The overall objective of state policy in the IT sector is to create a group of large, internationalized companies, preferably though not necessarily state-owned, which can (a) satisfy most of the demand for IT products and services from government agencies and major state-owned enterprises; (b) generate their own patented technology; and (c) play a leading or dominant role in setting international technology standards. This objective marks something of a tactical shift from the key objective of the late 1990s, which was to create indigenous substitutes for foreign technologies such as Intel's computer processor chips or the Windows operating system. While the effort to create core technology substitutes continues, it has become relatively less important, as the impressive success of a handful of Chinese hardware companies—notably Legend Computer (now Lenovo) in the personal-computer arena and Huawei Technologies in telecom network equipment—showed that China could produce highly competitive IT companies even without core technology substitutes.

Since 2000, the major emphasis of state IT policy therefore moved away from pushing state-supported labs to duplicate preexisting core technologies, and toward supporting companies that promise sufficient economies of scale and internationalization of outlook to be able to build significant in-house research and development (R&D) capabilities. At the same time, an extra layer of support was created for IT sectors in which China's domestic capacity was limited: semiconductors and software. In semiconductors, the initial aim was simply to reduce China's reliance on imported integrated circuits, or ICs. (For several years ICs have been one of China's top two import items—along with crude oil—accounting for around 10 percent of total import value.) In the longer term, the government would like to create indigenous semiconductor design firms, since design is the highest-value part of the IC manufacturing chain. In software, the main emphasis is to replicate India's success as a provider of software outsourcing services to Western and Japanese companies. As will be shown later, these state objectives do not necessarily coincide with the commercial objectives of individual Chinese firms.

2.2.1 Government policies and institutional mechanisms

The principal IT policy-setting bodies are the Ministry of Information Industry (MII), the Ministry of Science and Technology (MOST), and the State Commission on Science and Technology for National Defense (COSTIND), the latter focusing mainly on military applications of IT.[3] In the civilian sector, MII is by far the most important bureaucratic actor. As described later in Section 2.3, both MII and MOST operate research institutes whose main purpose is to gather and analyze technical information and market data. In addition, MII has regulatory authority over the telecoms and internet industries, and uses this power to restrict foreign competition in various market sectors and to promote the interests of domestic equipment vendors.

Government-sponsored research on the development of new products and standards is mainly carried out by research institutes under CAS and at various universities, of which Beijing University and Tsinghua University (both located in Beijing) are the most important. Beijing's status as the center of IT research in China is further enhanced by the Zhongguancun High-Tech Park, a sprawling industrial zone near Beijing and Tsinghua Universities which is earmarked for the development of IT-related firms. With the notable exception of the major telecoms equipment vendors such as Huawei, most of China's prominent IT firms are based in Beijing.

The broad outlines of sector-specific government IT policies are sketched out here; further details are given in Section 2.4. In telecoms and the internet, government aims are mainly accomplished through MII's regulatory activities, the two principal features of which are (1) severe restrictions on foreign ownership of telecoms operating companies and (2) informal market-share limits for foreign telecoms network equipment producers. The manufacturing of much IT equipment (e.g., personal computers and mobile phone handsets) is, like most manufacturing in China, very lightly regulated.

In semiconductors, policy from 2000 to 2004 was guided by State Council Document 18, issued in 2000, which provided an array of tax and other incentives for domestic semiconductor manufacturers.[4] These policies contributed to sparking substantial semiconductor manufacturing investments in China. In addition, MOST directly financed more than 100 start-up IC design firms. The incentive programs were fiercely criticized by foreign semiconductor firms who argued that they were not compliant with China's obligations in the World Trade Organization (WTO), and these were cancelled in 2005.

Software firms also received incentives under Document 18—although they failed to produce the same positive effects as in the semiconductor industry. These were also cancelled in 2005. Unlike the IT hardware and semiconductor industries, which are viewed as strategic manufacturing industries and as such enjoy widespread support throughout the government, software is something of a stepchild. It does not enjoy the benefit of systematic industrial development policies comparable to the Software Technology Parks of

India (STPI) program, which played a large role in the development of the Indian software industry.[i] Neither MII nor MOST appears to view software as a major priority. Where software outsourcing has begun to thrive—most notably in the northeast city of Dalian—it is mainly the result of local initiatives, not support at the national level.

2.2.2 Funding of IT research

Information technology's share of government research funding has steadily shrunk over the past decade, reflecting the increased commercialization of IT in China. It appears to be a government goal to have a greater proportion of IT research conducted by enterprises, thereby freeing scarce government research funds for application to less commercially advanced sectors.

In 2004, according to data from MOST, "promotion of industrial development and technology" accounted for around half of all government research funding (except in basic research where the share is much smaller) and one-quarter to one-third of the manpower allocation (see Figure 2.2). These figures clearly show the industrial orientation of China's overall research program, but they overstate the research commitment to IT specifically. In MOST's "Key technologies R&D program"—the single biggest conduit of government IT research funding—funds for IT fell by more than half between 1998 and 2004, and IT's share of the total dropped from two-thirds to one-fifth (see Figure 2.3). During the same period, funding for research in agriculture and biotechnology rose sevenfold and now accounts for half of Key technologies program spending.[ii] Assuming a comparable share for IT in the 863 program (a government fund established in

[i] Many Chinese cities have set up software industrial parks, but these are mainly local initiatives. Moreover, they are in essence real-estate development projects and so differ fundamentally from the STPI scheme, which despite its name was mainly a mechanism for creating an alternative international telecoms system for India-based software companies, enabling those firms to bypass the inefficient state-run telecoms firms. (Author site visits in Bangalore, Gurgaon, Kolkata, Hyderabad, Beijing, Shenzhen, and Guangzhou, 1999–03.)

[ii] Another research funding agency is the National Natural Science Foundation of China (NNSF), an independent ministerial-level agency that distributes a portion of 863 and 973 program money, as well as other funds allocated by the government, via a competitive grant system.

Figure 2.2 Industrial development and technology share of total government research funding and staffing, 2005.

	Funds		Staff	
	Rmb m	% of total	man-years	% of total
Total industrial development/ tech share of which:	18,179	56.9	24,494	25.8
Key technologies R&D program	12,902	71.0	7,586	31.0
863 program	5,044	27.7	14,964	61.1
Basic research programs (973 and others)	232	1.3	1,943	7.9

Source: *China Statistical Yearbook on Science and Technology 2006*, p. 277.

Figure 2.3 Funding for IT research under the Key Technologies R&D Program.

	1998	2001	2004
High-tech industry, rmb m	700.0	374.0	300.0
High-tech industry, % of total program S&T expenditure	67.5	35.2	20.5

Source: *China Statistical Yearbook on Science and Technology 2005*, p. 354.

March 1986 to finance science and technology R&D), and a lesser share in basic research, it appears that total direct government support for IT research was probably not much more than RMB 1 billion (USD 125 million) in 2004. This is a tiny fraction of the RMB 27.7 billion (USD 3.4 billion) spent on R&D by the IT industry in 2005.[5]

This estimate, however, understates the true level of government support for IT research. Although IT research has been pushed into enterprises, major IT enterprises receive substantial government support which is difficult to quantify in a systematic way. The most obvious example is Huawei Technologies, a telecoms equipment firm that reportedly obtained soft loans at an early stage of its development in the 1990s, and in 2004 received an RMB 10 billion (USD 1.25 billion) line of credit from the China Development Bank, a policy lending institution of the central government. While this sort of subsidy is not research support per se, it is clearly targeted at the relatively small number of companies, such as Huawei, which have invested heavily in R&D.

As direct government spending on IT research has declined, R&D spending by IT companies has increased, from RMB 11.6 billion (USD 1.4 billion) in 2001 to RMB 27.7 billion (USD 3.4 billion) in 2005. In 2005, the IT industry accounted for 22 percent of all industry spending on R&D in China. However, the apparently large amount spent by industry on R&D almost certainly overstates the impact of this research. Enterprise R&D is highly fragmented, lacks economies of scale, and in most cases is duplicative or focused on minute incremental improvements to improve the salability of products; much of it focuses on relatively low value products and on consumer electronics. Only a handful of the very largest companies devote serious effort to producing real technological innovation.[6]

2.2.3 Standards and intellectual property rights

Information technology depends heavily on two pieces of institutional software: IPR and standards. Explicit and implicit government policies on standards and IPR underpin the overall IT policy stance.

The generally accepted view in the United States and Europe is that IPR encourages innovation by enabling innovators (whether in technology, process, distribution, or branding) to profit from their inventions. Standards enable wide diffusion of IT by ensuring that systems and products from different producers can function together.

Implicit in Chinese IT policy is an almost completely opposite view of the roles of IPR and standards. Protection of IPR has long been seen as a constraint on technology diffusion (because it makes products more expensive). Proprietary technology standards are increasingly seen as a potential way for China to strengthen the market position of domestic firms and reduce the influence of foreign ones. Neither of these views is absolute or fixed: many in the Chinese government advocate the development of a stronger IPR regime, and acceptance of non-proprietary international standards is widespread. Nonetheless, the ideas of IPR as a constraint on, and proprietary standards as a crutch for, domestic enterprises continue to enjoy wide appeal and exert significant influence on IT policy.

The starting point for Chinese economic policy since the 1980s has been that technology must be disseminated widely and rapidly, and at the lowest possible cost. In practice this has meant that intellectual property rights have been poorly enforced, since the protection of IPR holders and the payment of royalties and license fees are seen to conflict with the goals of low-cost and rapid technology diffusion. Foreign software firms estimate that more than 90 percent of software used in China is pirated; until 2006 it was customary for Chinese vendors to ship computers without an operating system, on the assumption that buyers would install pirated copies of Windows.

Conversely, the government has tended to view standards as mechanism not for diffusion but for revenue creation. Most standards are open and non-proprietary: that is, they specify standard ways of doing things but do not require the payment of a license fee. China has embraced this standardization process—for instance, encouraging its firms to meet International Organization for Standardization (ISO) 9000 standards on manufacturing quality and ISO14000 standards on environmental management.

In a few cases, however, standards and IPR overlap because a core proprietary technology developed by a particular firm or consortium is adopted as an industry-wide standard. In such cases, manufacturers using the core technology must pay license fees to the developers. Prominent examples are the Moving Picture Experts Group (MPEG) standard for compression of digital video files (controlled by a consortium of European and Japanese firms), used in the production of Digital Video Disc (DVD) players; and the Global System for Mobile Communications (GSM) and Code-Division Multiple Access (CDMA) standards for mobile-phone networks.

The Chinese government has recently seized on the development of proprietary standards as a mechanism by which Chinese firms can earn royalty income from domestic inventions—in effect, a narrow IPR regime which can be made to benefit only domestic and not foreign firms. (Developed-country governments and companies also frequently promote proprietary standards, so China is by no means an outlier in this regard.) The interest in proprietary standards intensified after 2000, when it became clear that efforts to achieve technological independence by creating a Chinese semiconductor design (to replace Intel's chips) and a Chinese computer operating system (to replace Windows) had failed. But the standards push has achieved no notable results. China has 21,000 national standards and has adopted about 30 international standards. Less than 10 IT standards have any Chinese intellectual property content and no more than five products are sold using Chinese-developed standards. There are no Chinese standards licensed to producers elsewhere, or involved in cross-licensing arrangements.[7] Two widely publicized Chinese standards-setting efforts are examined later in Section 2.4.4.

The impact of China's IPR and standards policies on innovative capacity is mixed. The crucial issues are the extent to which IPR and standards policies enhance technology diffusion and increase rewards for innovators. A body of research suggests that IT consumers are key drivers of innovation, and that IT consumers reap far more of the economic benefit of innovation than do the producers.[8]

One implication of this view is China's lax IPR enforcement which—though decried by foreign patent-holders—has bolstered China's long-run innovative capacity by producing a large body of both individual and corporate IT consumers whose demands will drive innovation in the future, in ways that are now difficult to predict. At the same time, lack of patent protection and rules that assign most of the financial benefit of patent rights to corporations and government research institutes, rather than to individual inventors, means that the financial incentives for innovation are poor.

The obsession with proprietary standards, meanwhile, neither promotes technology diffusion nor increases the financial rewards for innovators in any substantial or sustainable way. It is merely an effort to create a small but obvious revenue stream which can easily be captured by the state or its agents. Far more important will be China's openness to international non-proprietary standards. A high degree of openness will enable innovative Chinese firms to make incremental improvements, which can be swiftly marketed in both domestic and international markets. (An important side-effect will be that as their market power grows, Chinese manufacturers will gain a much larger say in determining the content of international standards.) Insisting on a duplicative body of idiosyncratic national standards will impede innovation by making it more difficult for Chinese firms to cater to both the fast-growing but low-value domestic market and slow-growing but higher-value international markets.

2.3 Information technology-oriented research institutes

A large number of government-funded research institutes are involved in IT. Some mainly collect and disseminate data; others actually employ scientists, engineers, and students to conduct basic research and engage in product development. Research in this area tends to be concentrated in Beijing, although Shanghai is a secondary center. There are three main types of civilian institutes:

1. *Ministry-sponsored research institutes*. These tend to focus their efforts on the general collection and dissemination of information, rather than on practical research aimed at product development. The two most prominent are the MII's Center for Communications Industry Development (CCID) and MOST's Institute of Scientific and Technical Information of China (ISTIC). The main mission of both is to support the development of IT policy through analysis and research. The CCID, the larger and better-known of the two, publishes authoritative reports on the telecoms, electronics, and internet industries. It also runs eight research centers and 41 enterprise units (of which two are listed companies), which receive funding under various national R&D programs. Some of these units are involved in product development but most focus on technical training, information gathering, and analysis. The MII also runs a separate institute, the China Software and Integrated Circuit Promotion Center (CSIP), which performs functions similar to CCID's in the software and semiconductor industries.
2. *Research institutes under CAS*. Three CAS institutes do serious R&D in information technology: the Institute of Computing Technology (ICT, the parent organization of Lenovo), the Institute of Semiconductors, and the Institute of Software. These are described in more detail later.
3. *Research institutes under major universities*. The most prominent of these in the IT area are Tsinghua University's Research Institute of Information Technology (RIIT), and Beijing University's Institute of Computer Science and Technology (ICST).

Both CAS and university-sponsored research institutes operate under a clear mandate to develop commercially viable products that can be spun off into companies. Examples of the institutes' spin-off companies include Lenovo, China's biggest computer firm (spun off from the Chinese Academy of Sciences); Founder, China's largest developer of software for the printing and publishing industries (spun off from Beijing University); and Neusoft, China's biggest software company, which was originally a creation of the Northeast University's software development center in Shenyang.

On the whole, CAS research institutes are larger and better funded than their university counterparts (see Figure 2.4). Each of the CAS institutes has around 500 full-time staff and a large array of international cooperation projects and spin-off companies. Each derives about a third of its funding from CAS, another third from grant funding from the NNSF, and the rest from other external sources, which include income from spin-off companies.

The grandfather of IT research institutes in China is CAS Institute of Computing Technology, which was founded in the 1950s and has trained much of the key staff at other IT institutes at CAS and elsewhere. It has spun off more than 30 companies of which the most prominent are Lenovo, Dawning Information Industry Co., and BLX IC Design. In addition to its well-publicized listed arm that acquired IBM's personal-computer business in 2005 to become the world's third largest PC vendor, Lenovo's unlisted parent company produces supercomputers designed by ICT engineers which have been used in China's space program. Dawning markets another ICT-developed line of high-end servers and supercomputers, used for enterprise networks, gene sequencing and satellite imaging; government agencies and large state companies are the major customers. The Longxin microprocessor is sold by BLX IC.

The ICT has seven major laboratories, two of which are designated National Engineering Research Centers (NERC), entitling them to higher levels of central-government funding. These are the NERC for Intelligent Computer Systems and the NERC for High-Performance

Figure 2.4 Chinese Academy of Sciences enterprises.

	1995	2000	2004	2005
Income, rmb bn	5.4	36.8	46.7	109.8
Profit, rmb bn	0.4	2.0	9.4	10.5

Source: CAS Audit office, http://jianshen.cashq.ac.cn/info_www/news/ detailnewsb.asp?infoNo=1049.

Computers, both of which work on the development of supercomputing systems. Five other laboratories work on developing smaller scale central processing units, software applications, and internet and wireless telephony systems. The ICT has a long roster of international cooperation projects with Western universities and companies including MIT, Princeton, IBM, Motorola, and Microsoft.

The Institute of Software runs a large number of research projects, some with the assistance of foreign partners including IBM, Motorola, Microsoft, and Japan's NEC. It has spun off five companies, including a joint-venture with NEC and two well-known firms, Sinosoft and Red Flag. These two companies respectively represent what might be termed the "opportunistic" and the "national strategic" varieties of spin-off firms. Sinosoft is a diversified software firm which has produced systems for highway departments, postal bureaus, and museums; it also has a training arm. Like Lenovo (another CAS spin-off) it focuses on developing useful products adapted to the Chinese environment, rather than on creating innovations that are globally relevant. Red Flag, meanwhile, is a relic of the 1990s dream of creating an indigenous Chinese-language, Linux-based operating system that would enable Chinese computer users to free themselves from their expensive reliance on the foreign Windows system.

The Institute of Semiconductors, despite its name, is mainly notable for its achievements in optical electronics; its main innovations have been a variety of laser technologies including some for use in products such as DVD drives, players, and burners. It has produced nine spin-off firms, most of which are fairly small and which mainly specialize in optical products such as light-emitting diodes (LED) for traffic lights and electronic displays, laser diodes for use in surgical procedures, and fiber-optic communications equipment. Only one spin-off firm, the Beijing Video and Audio Microelectronics Design Center, is significantly involved in the design of integrated circuits.

On the university side, the most important institute is Tsinghua's RIIT, which employs 300–400 professorial and engineering staff split among 10 research centers, three of which are designated National Engineering Research Centers. It focuses on third-generation mobile telephony, wireless networks, and computer processor chip design. A substantial part of its research relates to the creation of hardware, software, and networks for major state enterprises and government agencies. Recent projects have included the development of an e-government system for the Beijing municipal government, an enterprise resource planning (ERP) system for the China National Petroleum Corp., and a data management system for China Mobile Corp.

Beijing University's ICST has 100 full-time staff researchers, plus another 100 masters and doctoral students who contribute some time. Its major achievements have been in the creation of systems for electronic editing and typesetting of Chinese text, including specialized programs for book and newspaper publishing. At present it focuses on digital media, security, and electronic commerce programs. It has no spin-off companies and is not involved in any cooperative ventures with foreign partners; instead all of its research feeds into product development at Founder, with which it has close ties. It is, in effect, a major R&D lab for Founder.

Outside of Beijing, two significant university-linked IT research institutes are the Shandong University School of Science and Technology, and the Xian-based Xidian University (*Xian Dianzi Keji Daxue*, literally Xian University of Electronic Science and Technology). The latter runs a network-security lab that reports to the Ministry of Education, and a wireless communication lab under MII.

2.4 Major IT production sectors in China

The list of China's largest IT manufacturing firms, as compiled by MII, is somewhat misleading (see Figure 2.5). Most of the companies on the list rely heavily on sales of household appliances such as refrigerators and air conditioners, or mature consumer electronics items such as color TVs. Profit margins and R&D expenditure are relatively low, and there is no sign of an increasing trend in either (see Figure 2.6).[iii] This section examines three sectors of IT production where Chinese innovation is either already emerging or has good prospects of doing so: telecoms (including the internet), semiconductors, and software.

[iii] Average R&D spending of the top 100 Chinese IT companies, 3.7 percent of revenues, compares to an 8.2 percent average R&D spending by the top 33 global technology hardware and equipment manufacturers. See "The R&D scoreboard, 2006," Department of Trade and Industry of the United Kingdom, http://www.innovation.gov.uk/rd_scoreboard/downloads/2006_rd_scoreboard_data.pdf.

Figure 2.5 China's top ten IT manufacturing firms by revenue, 2005: key indicators.

Rank	Company	Revenue USD m	Profit USD m	Profit margin (%)	Exports USD m	Exports/ revenue	R&D exp USD m	R&D/ revenue	Main products
1	Lenovo Group	13,210	251	1.9%	1,407	10.6%	183	1.4%	Computers, printers, mobile phones
2	Hai'er Group	12,691	161	1.3%	1,311	10.3%	557	4.4%	Color TVs, refrigerators, air conditioners
3	BOE Group	6,693	(8)	−0.1%	1,017	15.2%	113	1.7%	LCDs, color monitors, color kinescopes
4	TCL Group	6,367	(144)	−2.3%	1,170	18.4%	238	3.7%	Color TVs, mobile phones, computers, air conditioners
5	Huawei Technologies	5,735	629	11.0%	2,117	36.9%	580	10.1%	Fixed and mobile telecommunications network equipment
6	Midea Group	5,189	95	1.8%	1,777	34.3%	163	3.1%	Air conditioners, microwaves, household appliances
7	Hisense Group	4,075	76	1.9%	452	11.1%	175	4.3%	Color TVs, air conditioners, refrigerators, switching equipment
8	SVA Group	3,578	(55)	−1.6%	1,900	53.1%	123	3.4%	Mobile phones, color TVs, color kinescopes
9	Panda Group	3,434	76	2.2%	1,800	52.4%	42	1.2%	Mobile phones, wireless stations, switching equipment, color TVs
10	Founder Group	3,160	104	3.3%	106	3.4%	155	4.9%	Electronic publication systems and software, computers

Source: Ministry of Information Industry (MII) (see endnote 4), Dragonomics Research.

Figure 2.6 Profit margins and R&D expenditure at China's top 100 domestic IT manufacturing firms.

	2002	2003	2004	2005
Profit margin, %*	4.16	4.27	4.08	2.56
R&D/revenue, %	3.80	4.08	3.82	3.73

Note: * Profit as % of revenue.

Source: Ministry of Information Industry (MII), Dragonomics Research.

It concludes with a brief discussion of two prominent efforts to establish proprietary domestic IT standards.

2.4.1 Telecoms and the internet

The development of China's telecommunications industry is one of the most impressive achievements of the past 15 years. In 1990 China had a telecommunications system not much better than that in most developing countries. Within a decade it had a first-rate telecoms system and by far the highest mobile phone and internet penetrations of any low-income country. In absolute terms, China has more mobile phone users than any other country, and ranks second in internet users behind the United States.

Two main factors account for this spectacular growth. The first is the regulatory regime, a classic example of "managed competition." This competition occurred at two levels. At the national level, competition was created in 1994 with the establishment of a second telecom service provider (China Unicom) to vie with the Directorate-General of Communications (DGT). The DGT was subsequently corporatized as China Telecom. A similar duopoly was then created for mobile phone services.

More important was the decentralization of management of local loops. Provincial governments, which compete against each other for investment and growth, thus also competed against each other in the creation of telecoms infrastructure—both basic networks and value-added services. By the mid-1990s thousands of independent telecommunications providers offered services such as paging, internet access, and electronic data interchange. Local governments had great incentives to improve telecoms networks because doing so helped attract foreign direct investment and boosted economic growth—two key criteria for promotion of government officials.

A second factor aiding China's telecoms growth was the emergence of indigenous telecoms equipment makers. These firms developed adequate technology and sold it at a far lower cost than their foreign competitors. This again made it far more economical for localities—especially outside the major cities where foreign vendors concentrated their efforts—to build networks. By 2005, domestic firms controlled 45 percent of the national market for telecoms network equipment (they dominated the fixed-line sector with a share of 75 percent, but in the more technologically advanced mobile network sector their share was only eight percent).[9] The leading domestic telecoms network vendor, Huawei Technologies, is one of the few large Chinese firms to have demonstrated real innovative capacity and is considered separately later.

Growth in internet usage has also been impressive. At the end of 2005, China ranked second in the world with around 100 million internet users.[iv] These usage figures mainly reflect the large size of China's urban population and the success of the Chinese economy at making computers and telecommunications cheaply available. This is a substantial achievement but not in itself indicative of innovation. By a more relevant measure of how large a role the internet plays in the national economy—the number of internet hosts—China ranks thirty-ninth worldwide, behind Mexico, Turkey, and South Africa (see Figure 2.7).

China's high internet usage rate reinforces the point that it has diffused IT more broadly than most countries of comparable average income. But the relationship between IT adoption and IT innovation is not straightforward. As noted above, widespread technology adoption in the long run may create an environment conducive to innovation, by generating a large pool of consumers whose demands drive producer innovation. This effect, however, is unlikely to emerge on a large scale until the aggregate spending power of these consumers is large enough to make incremental improvements highly profitable.

[iv] The Ministry of Information Industry estimates 111 million users; the most recent edition (June 2006) of the bi-annual survey by the China National Network Information Center, which has consistently published higher estimates than MII, reported 123 million.

Figure 2.7 Internet hosts by country, 2004.

Rank	Country	No of hosts, m	% of global total
1	United States	195.1	72.9
2	Japan	16.4	6.1
3	South Korea	5.4	2.0
4	Netherlands	5.4	2.0
5	United Kingdom	4.2	1.6
17	Finland	1.2	0.4
39	**China**	**0.16**	**0.1**
40	India	0.14	0.1

Source: International Telecommunication Union (ITU), World Telecommunications Indicators 2005.

China has not yet reached this point: the volume of potential users is large but average spending power is low. In such a market the most effective business strategy is to provide generic services to as many customers as possible at the lowest cost.

At present, Chinese internet firms commoditize (produce at high volume and low cost) technologies developed first in richer markets. Innovations mainly involve ways of circumventing market inefficiencies such as the difficulty in generating revenue and collecting payments in a country without credit cards. The first wave of Chinese internet firms to list on international stock markets (notably Sohu, Sina, and Netease) were portals offering services similar to those of Yahoo: email, search, news, and chat. Most of them foundered after listing because of a dearth of revenues. Their fortunes recovered in 2003 with the explosion of mobile phone short message services (SMS), which piggybacked on the portals' messaging systems and enabled them to collect fees from the mobile phone operators. More recent waves of Chinese internet firms have copied e-commerce business plans first developed in the United States. These include eBay-like auction sites, an online travel service (C-trip), similar to Expedia, and a Google-like search engine, Baidu. In many cases these Chinese firms have won a greater share of the Chinese market than the more innovative and better funded U.S. firms they are copying. This mainly reflects their ability to cater to local consumer taste—which reflects innovative capacity in marketing rather than in technology.

Huawei Technologies

The biggest, most international, and most analyzed of China's domestic telecoms firms is Huawei, a private firm established in 1988 by several former members of the logistics operation of the People's Liberation Army. Huawei has achieved a certain mystique because of its success, the army links of its founders, and because its chairman Ren Zhengfei never gives interviews. The balance of evidence suggests that Huawei is what it claims to be: a private telecoms equipment firm that has prospered by dogged marketing and a strategic focus on R&D spending. The army links of its founders have never been shown to be more than incidental.

The consensus among international telecom firms is that Huawei got its start by reverse-engineering imported telecoms switching chips, and its growth in the late 1990s was allegedly smoothed by low-interest loans from the government. But Huawei's foreign competitors now admit the firm's technology is internationally competitive and includes many incremental innovations developed in-house.

Huawei is one of the very few Chinese companies that can plausibly claim to be R&D driven. It has spent at least 10 percent of revenues on R&D since 1993. Its international competitiveness is underscored by its success in overseas markets. In 2001, international sales accounted for just 11 percent of Huawei's revenues of USD 2.2 billion. In 2005, that figure was 58 percent (of a much higher total revenue of USD 8.2 billion). Most of that reflects sales of mobile networks in developing-country markets in Africa and elsewhere. But it has also won supply contracts for telecoms firms in the United Kingdom and the Netherlands. In 2005, Huawei ranked a close third behind Ericsson and Nokia in the number of contracts for new mobile networks globally, although the value of its contracts was less than 2 percent of the global total.[10]

Huawei casts light on China's innovation potential in three ways. First, it demonstrates how individual Chinese firms can overcome the general tendency toward commoditized, low-cost production. Second, its very uniqueness suggests that such firms are likely to be the exception rather than the rule for some years to come. Even the second-biggest Chinese telecoms equipment maker, state-owned ZTE, lags far behind Huawel in foreign markets (international sales in 2005 were

USD 951 million, or 36 percent of total revenues).[11] It is not clear how easily the Huawei model can be replicated in any segment of China's IT industry.

Third, and most relevant to current government policy, Huawei has flourished without controlling a proprietary standard. Instead, it has worked with internationally accepted technology standards (notably the GSM and CDMA standards for mobile telephony) but innovated in order to reduce the cost of technology and marginally to improve its functions. Like ZTE and other Chinese hardware vendors, Huawei appears lukewarm toward the government's drive to create proprietary domestic technology standards.

A final point is that while Huawei's innovative capacity is impressive by domestic standards, it is still modest in comparison to multinationals and even the mainland operations of Taiwanese firms. Of the 616 U.S. utility patents filed by China-based firms between 1997 and 2004, only 11 were filed by domestic firms; Huawei had six of these. By contrast, firms controlled by Taiwanese, Hong Kong, or other overseas Chinese interests filed 503, more than 80 percent of the total.[12]

2.4.2 Semiconductors

China's semiconductor or integrated circuit industry has notched up impressive progress since 2000, partly thanks to generous tax incentives. Nonetheless, China remains firmly locked at the very lowest level of the semiconductor value chain: China is far more important as a consumer than as a producer of integrated circuits. According to the China Semiconductor Industry Association (CSIA), in 2005 China accounted for 21 percent of world IC consumption value, but only 3.7 percent of world production value.[13]

The semiconductor value chain consists of three major elements. First, and most lucrative, is chip design, which is extremely skill- and R&D-intensive and is controlled by a handful of firms of which the largest is Intel. Second is chip manufacture, which is highly capital intensive; a semiconductor plant (or "fab") typically requires investment of USD 1–3 billion. The major chip design firms such as Intel, IBM, and Texas Instruments generally operate high-end manufacturing plants themselves, but outsource the production of lower-value chips to contract manufacturers, many of which are in Taiwan. The final and lowest-value stage is the assembly, testing, and packaging of pre-manufactured chips (a process generally referred to as "assembly and test"). A rough idea of the relative value of the three stages of IC production can be gleaned from these figures: in 2005, Intel's semiconductor revenues were USD 35.7 billion, while those of Taiwan Semiconductor Manufacturing Corporation (TSMC), the world's largest contract manufacturer, were USD 8.3 billion. The entire Chinese semiconductor industry generated revenues of USD 9.3 billion, only slightly more than TSMC and a quarter of Intel's figure.[14]

Until 2002, virtually all of China's semiconductor industry was in assembly and test. The past three years have seen rapid growth in contract manufacturing (mainly by Taiwanese firms) and in design. Assembly and test still accounts for the biggest share of China's industry, but that share is shrinking (see Figure 2.8). Nonetheless, China's reliance on imported semiconductors has continued to rise. The average value of semiconductors produced in China has barely crept up (to 6 cents in 2005), whereas the average value of imported semiconductors has more than doubled, to USD 1.08 in 2005 (see Figure 2.9).

This semiconductor deficit is the flip side of China's prowess as a consumer electronics assembler. All electronic products contain chips; China in general lacks the capacity to design or manufacture these chips, so it must import them. A key government goal is for China to design and manufacture its own chips, reducing import dependence and increasing the local value-added of electronics exports.

Figure 2.8 Structure of China's semiconductor industry revenues.

	2004		**2005**		**2004–05**
	Value rmb bn	**% of total**	**Value rmb bn**	**% of total**	**% growth**
Design	8.1	14.9	13.1	17.4	60.8
Manufacture	18.1	33.2	28.0	37.3	54.5
Assembly/test	28.3	51.8	34.0	45.3	20.3
Total	**54.5**	**100.0**	**75.1**	**100.0**	**37.7**

Source: China Semiconductor Industry Association (CSIA), via Xinhua, "Analysis of China Semiconductor Industry", endnote 13.

Figure 2.9 China's semiconductor trade balance.

Imports	**2000**	**2005**
Import volume (bn units)	20.5	75.4
Import value (USD bn)	9.5	81.0
Average value per unit, USD	0.46	1.08
Exports	**2000**	**2005**
Export volume (bn units)	40.4	216.1
Export value (USD bn)	2.0	13.8
Average value per unit, USD	0.05	0.06
Balance	**2000**	**2005**
Net volume (bn units)	19.9	140.7
Net value (USD bn)	(7.5)	(67.3)
Ratio of import/export price	9.49	16.80

Source: Ministry of Information Industry, http://www.mii.gov.cn/art/2006/03/15/art_62_8309.html.

To facilitate this aim, the State Council in June 2000 issued its Document 18, "Policies to Encourage the Development of the Software and IC Industries." This policy simplified the approval procedures for semiconductor investments and lowered the value-added tax (VAT) on domestically manufactured semiconductors to six percent. Start-up semiconductor firms also received substantial funding from MOST. Between 2000 and 2003, MOST financed the creation of more than 100 IC design centers. In 2002, another State Council decision announced incentives for venture capital in the IC industry, most notably the promise that companies in this sector would get quicker approval for stock-market listings. These national-level incentives were supplemented by local programs of tax breaks and direct subsidies, most significantly in Shanghai and in nearby Suzhou and Kunshan. Localities may also have given additional, unpublished subsidies to preferred companies.[15]

The preferences extended in Document 18 were officially abolished in April 2005, following vociferous complaints from foreign semiconductor firms that they constituted subsidies illegal under

WTO rules. With the spectacular growth of China's electronics industry, such subsidies are no longer really necessary, since market forces will drive substantial further investment in local semiconductor production. Nonetheless, the central government is considering a new set of preferences to replace the abandoned Document 18 ones. These may include:

- An Integrated Circuit Development Fund, under which the Ministry of Finance would distribute up to RMB 1 billion each year to IC design firms;
- Extending tax holidays on IC firms from two years to as long as 10 years;
- A one percentage-point discount on the interest rate for new loans in the IC sector.

The impact of either the old policies or the forthcoming ones on the development of innovative capacity is far from clear. Between 2000 and 2005, Japanese firms registered 25,000 IC-related patents in the United States; during the same period Taiwanese firms registered 8000 (up from 2000 in the preceding five-year period) and Chinese firms registered less than 500.[16] While Chinese design firms can be expected to accelerate their pace of innovation and overseas patent filings, this will do them little practical good unless they can protect their patents from copying at home. The time from chip design to commercial-scale production typically ranges from three to five years, leaving ample time for designs to be stolen and copied. The main story of China's semiconductor industry over the next decade is likely to be one of capacity-building, rather than the development of indigenous innovation.

2.4.3 Software

Software is the laggard in China's otherwise impressive IT industry development. Nine of the top ten "software" companies in China are really hardware companies that engage in software development as an offshoot of their main business. The biggest pure software firm, Shenyang-based Neusoft, generated revenue of USD 281 million in 2005, which would put it a distant sixth in India. Many large Chinese firms derive much of their revenue from system integration, rather

Figure 2.10 China's top ten pure software firms by revenue, 2005.

Rank	Company	Revenue		% of revenues from system integration
		USD m	% change	
1	Neusoft	281	28.0%	54
2	Microsoft (China)	235	na	na
3	CS&S	203	17.9%	61
4	CVIC	184	−56.9%	64
5	Baosight	163	20.7%	40
6	Oracle (Beijing)	161	27.7%	29
7	UFSoft	122	12.8%	2
8	SAP	98	27.6%	na
9	Kingdee	71	12.0%	13
10	Danda Soft	42	−15.4%	na

Note: Excludes online game developers.

Source: China Center for Information Industry Development (CCID), Dragonomics Research.

than from original software development.[v] The top ten list of pure software firms also includes local development centers of three multinationals: Microsoft, Oracle, and SAP (see Figure 2.10). The biggest domestic makers of off-the-shelf proprietary software are Founder (publishing software), UFSoft, (enterprise resource planning or ERP systems), and Kingdee (accounting software). Founder is the largest and relies heavily on hardware sales. The other two are small, and Kingdee has recently suffered sharp reverses. The biggest software outsourcing firms are Neusoft and CS&S; in each case outsourcing accounts for around 15 percent of total revenue.

Thus, China has neither an outsourcing industry remotely comparable to India's, nor an indigenous software industry producing original packaged software, or enterprise solutions. The Chinese government recognized this deficiency in the late 1990s, and the State Council's Document 18 provided tax and other incentives for software firms along with the previously discussed incentives for semiconductor

[v] System integration is simply the work of making different software programs run together smoothly in a single system or network. It can involve the coding of patches or utilities that enable different software programs to run together, but the amount of programming required is many times less than in the development of an original program, or in outsourcing activity.

firms. Local governments were encouraged to set up software development parks, in an attempt to emulate the Software Technology Parks of India scheme.

The Document 18 incentives produced little effect in software. One problem is that incentive schemes designed to reduce the capital cost of start-up manufacturing industries provide little benefit to service industries with low capital requirements. But two deeper reasons are more important: lack of market opportunity and intellectual property theft. India's software outsourcing industry developed in part because the domestic software market was so tiny. With little money to be made at home, Indian firms had a huge incentive to shop their services abroad. Chinese firms, meanwhile, enjoy a domestic market of government offices and state enterprises with enormous system-integration requirements—a downstream effect of China's high rate of technology diffusion. Chinese firms thus have little incentive to pursue outsourcing contracts.[17]

China's extremely poor IPR protection affects domestic software developers in two ways. It makes the development of packaged software virtually impossible, since any new program will be instantly pirated. It also makes outsourcing difficult because Chinese firms face a high credibility threshold when it comes to convincing potential offshore clients that proprietary software code will be safe in their hands.

In this bleak landscape three potential pathways to innovation stand out:

1. *Research and development centers by foreign software and outsourcing firms.* Foreign software firms are investing in Chinese development centers, attracted by the plentiful supply of low-cost software engineers and the relatively good telecommunications infrastructure. These centers initially provide localization of their standard products (such as Chinese-language interfaces), but a few are increasingly doing basic development work on new products for the global market. In addition, foreign outsourcing firms—including big Indian companies and also global technology consulting and outsourcing firms such as IBM and Accenture—have set up international outsourcing centers, principally in northeast China to serve the Japanese market. The northeastern city of Dalian is the center of this activity; Dalian claimed that Japan-oriented outsourcing revenues reached USD 375 million in 2004.[18]

2. *Embedded software*. Neusoft started providing embedded software for Alpine, a Japanese maker of car stereo systems, in the 1990s. It later set up a joint venture with a Chinese manufacturer of medical scanning equipment, again with the idea of providing a vehicle for sales of embedded software systems. This model makes sense given China's competitiveness as a manufacturer of consumer electronics. Yet few Chinese software firms have followed Neusoft's lead. China's main consumer electronics firms have chosen to keep their software development in-house, rather than outsourcing it. Similarly, most major Japanese electronics firms, including Sony, Matsushita, NEC, and Hitachi, have wholly owned R&D centers in China—mainly in Dalian—and prefer to keep their embedded software development in-house. As in outsourcing, China is likely to be a low-cost platform for the innovation of foreign firms; innovation by purely domestic firms will be difficult.

3. *Gaming software*. One market where domestic consumption could drive innovation, and where piracy concerns are minimal, is gaming software. Revenues from online gaming are growing rapidly and large multiplayer role-playing games are relatively resistant to piracy. An additional advantage is that such role-playing games are culturally specific. In the long run, domestically developed games should dominate the market, as has become the case in South Korea.

From a policy standpoint, it appears that the central government understands that earlier incentive policies are essentially useless. So its main strategy is to coax big foreign software firms into investing in joint ventures with local firms, rather than simply setting up wholly owned R&D centers.

An example of this approach was a memorandum of understanding signed by Microsoft and the National Development and Reform Commission (NDRC) in April 2006, shortly after Hu Jintao paid a visit to Microsoft boss Bill Gates in Seattle.[vi] Microsoft pledged to invest USD 100 million over five years in ventures involving Chinese

[vi] The headline value of the MOU, USD 3.7 billion, is misleading, since USD 3.5 billion of that amount represents a five-year hardware purchase commitment by Microsoft. Most likely this simply represents the expected value of OEM production of Microsoft hardware products (such as the Xbox gaming console and various computer peripherals), which would have occurred anyway, without the MOU.

software firms or their international subsidiaries. Microsoft also promised to place orders worth USD 100 million (again over five years) with Chinese firms for technical support and software development. Finally, it agreed to set up an "NDRC-Microsoft Software Innovation Center" to help promote innovation among Chinese software companies.[19]

Such agreements will probably not do much for local software innovation. The agreement plainly enables Microsoft to maintain complete control of its China operations (by, for instance, setting up joint ventures and then buying out the local partners). Most likely, the Microsoft agreement was a *quid pro quo* for a central government decree, issued just the month before, requiring that domestically made personal computers be pre-installed with legitimate operating-system software prior to sale. The benefits to Microsoft of capturing a larger share of the China operating-system market dwarf its stated commitments to software innovation in China.[vii]

2.4.4 Standards controversies

One important aspect of China's IT industry is the ongoing government effort to promote domestic proprietary standards, as discussed earlier in Section 2.2.3. The government has little interest in promoting general IPR protection as this would mainly benefit foreign firms. It can, however, enable domestic firms to pioneer core technologies that become proprietary international standards (from which the developers can earn royalties); but this in effect amounts to enforcing a very narrow IPR regime exclusively benefiting domestic firms. From a techno-nationalist standpoint the appeal of this strategy is obvious. In practice the results have been disappointing, as illustrated by two prominent recent standards efforts.

2.4.4.1 Wireless networking: The WAPI controversy

In 2003, China's MII announced that all wireless devices sold in China (such as notebook computers) would have to conform to the home-grown wireless application protocol interface (WAPI) standard,

[vii] For example, leading Chinese PC maker Lenovo's December 2005 agreement to pre-install Windows on all its computers is expected to generate annual revenue for Microsoft of USD 1.2 billion (although at least half of that represents revenues that Microsoft was earning anyway from IBM, whose PC business was acquired by Lenovo in early 2005).

developed by a company spun off from Xian University. Since WAPI technology was available only from Chinese vendors, this rule would have compelled foreign technology firms to license WAPI technology and, perhaps more important, reveal key elements of their own technology to the Chinese vendors in order to get the WAPI system to work properly. The MII's rule provoked strong protests from foreign firms, and Intel—which is committed to the dominant international wireless standard, Wi-fi[viii]—threatened to stop selling wireless-capable chips altogether in China. In the spring of 2004, after negotiations between the Chinese and U.S. governments, China agreed to postpone imposition of WAPI standard, and to work with international standards bodies to harmonize WAPI with other wireless standards.

In October 2005, however, the WAPI standard was submitted to ISO for approval as a global standard, in competition with a second-generation Wi-fi standard. In March 2006, ISO rejected WAPI, citing the technology developers' refusal to disclose their encryption algorithm, which made it impossible to assess its security claims. A working group also noted: "Attempts over the last two years by non-Chinese companies to procure any version of a WAPI device have failed."[20]

In January, before that decision was announced, MII and two other major government agencies announced that government departments would be required to buy only computer and telecoms equipment using the WAPI standard. Immediately after the decision of ISO, 22 Chinese telecoms and computer firms joined in a WAPI Alliance to promote the standard. It is not clear whether there are any products on the market actually using the WAPI standard, so the alliance may be a mechanism for creating products that will enable government agencies to comply with MII's mandate.

The WAPI story, although based on real technical concerns (proven security flaws in the original version of Wi-fi), quickly turned into one in which mercantilist concerns dominated. First, the government announced it would mandate the standard nationally without either (a) going through international standards bodies or (b) having conducted an open review process for domestic standards. Then, after being dragged into the international standards arena, the developer

[viii] The formal name of this standard is 802.11. A successor standard, 802.11i, which claims to address many of the security flaws of the earlier standard, is under development and is the main competition to WAPI. Both the 802.11 and 802.11i standards are known generically as "Wi-fi."

refused to disclose information necessary for the standards review, or to provide examples of products incorporating the standard. There is scant evidence that the standard incorporates any substantial innovation: WAPI is thus an example, not of domestic Chinese innovation capacity, but of a government-supported effort to favor domestic producers at the expense of foreign ones.[21]

2.4.4.2 Mobile telephony: The TD-SCDMA saga

The story of China's indigenous standard for "third-generation" or 3G mobile telephony (for telephones that can handle high-speed internet data transmissions) is more complex. The Chinese standard, known by its cumbersome acronym TD-SCDMA,[ix] had its origins in work done in the late 1990s in a research institute in central China. Later the German engineering firm Siemens took a significant stake in the standard-development project and it, along with various Chinese firms involved, would stand to benefit from any licensing income.

Unlike WAPI, TD-SCDMA has been accepted by the international telecoms standards body. By 2003 evidence was beginning to mount that the Chinese government would push hard to make TD-SCDMA either the exclusive or the preferred standard for China's 3G networks. As a result of negotiations with the U.S. government in 2004, China reaffirmed its commitment to standards neutrality, meaning that it could not adopt policies that would favor any one of the three internationally approved standards.

Subsequent government actions caused many outsiders to question the depth of China's commitment to standards neutrality. In particular, the government kept delaying the issuance of 3G licenses, apparently because it wanted to give more time for TD-SCDMA, which showed persistent shortcomings in trials. By the end of 2006, it was clear that Chinese mobile operators were not eager to receive the TD-SCDMA network license.[22] This underscores the point that Chinese government and industry do not necessarily share interests when it comes to standards. Many Chinese IT hardware producers are committed to production processes using international standards and are disinclined

[ix] Time division synchronous code division multiple access. The major competing 3G standards are CDMA2000, a successor to the CDMA standard developed by U.S. firm Qualcomm and WCDMA, the successor to the European GSM standard.

to risk new Chinese standards, especially since the major markets are still in the United States and Europe.

These examples suggest that the much-publicized Chinese domestic standards are not instances of commercially viable domestic innovation. Indeed, there is a serious risk that Chinese firms will waste time and money pursuing a chimerical goal of domestic proprietary standards. The most significant creation of Chinese IT intellectual property has been by Huawei, which has prospered by exploiting its mastery of international standards to generate highly competitive products that increasingly display incremental technological improvements. It is far more likely that IT innovation will emerge in China through incremental approaches by competitive firms like Huawei than by government attempts to rig standards-based monopolies.

2.5 Conclusions: Prospects of IT research in China

The overarching conclusion of the foregoing discussion is that China has not produced much technology innovation, but it has diffused technology throughout its economy far more rapidly than most other developing countries, including India. In the short term this constrains innovation because in a high-growth market with low average purchasing power, IT producers rationally pursue scale rather than innovation. In the medium- to long-term, widespread technology diffusion will create a large pool of technology consumers whose demands (once their average purchasing power rises sufficiently) will be an important impetus for innovation, just as American consumers are today.

At present, however, China has yet to demonstrate significant innovative capacity in any part of the IT industry—either in core technology, manufacturing process, or design. Indeed, it is in precisely the segment of the IT industry with the lowest manufacturing requirement and the highest need for innovation and intellectual capital—software—that China is weakest, notably contrasting with India. There are some exceptions, notably Huawei in the telecoms sector, that suggest that China may in the relatively near future begin to develop significant indigenous innovative capacity. This innovation is more likely to arise in companies providing products for business customers than in those serving the consumer market, because the higher profit margins (and lower branding and distribution-channel costs) mean that these firms have more funds available to invest in R&D.

More specifically, the following conclusions can be drawn about the nature and prospects of IT-related research and innovation in China:

Government-led research is adept at generating good-quality local versions of pre-existing technologies, but has produced little innovation. State universities and research institutes have spun off many IT companies, several of which have shown great proficiency at developing and marketing products based on technologies developed outside China. Almost none—with the possible exception of Chinese publishing software pioneer Founder—are notable for significant innovation. The most innovative Chinese IT company, Huawei, has always been private.

Research and development has been commercialized. Following on from the previous point, government funding for direct IT research has declined and has been significantly replaced by subsidies and incentives for companies. While the government will continue to conduct a significant amount of IT research in its own labs—especially in high-speed computing—most innovation is likely to take place in companies.

Commercial R&D in domestic firms is limited. China's comparative advantage in IT, as in other manufacturing sectors, is production at low cost and large scale, with low profit margins. Eventually, the cash flow of the biggest firms will be sufficient to finance significant R&D expenditure, even if profit margins remain low. But most firms are at least five years away from being able to finance enough R&D to generate significant innovation.

Future innovation will be driven more by consumer "pull" than industry or government "push." China's most impressive accomplishment in the IT field is its exceedingly high rate of technology diffusion. Up until now, and most likely for some years to come, this rapid diffusion comes at the expense of innovation, since it depends on a lax IPR regime that deprives innovators of financial reward. In the long run, however, the creation of an enormous domestic consumer base for IT products will provide an impetus for innovation, as consumers (both individual and corporate) demand more complex and sophisticated products. This effect will not materialize until average purchasing power rises substantially from current levels.

Foreign firms—especially those with Taiwanese or other overseas Chinese backing—will be the innovation leaders. Firms backed by Taiwanese or other overseas Chinese investors (including those founded by returnee Chinese with at least 10 years of overseas business experience) are by far the most innovative IT firms in China, accounting for over 80 percent of U.S. utility patent filings by China-based companies. Taiwanese firms in particular have mature design capacity, long-standing relations with customers in the United States and European markets, and experience in protecting intellectual property in a poor IPR legal framework (which they do mainly by employing trade-secret techniques rather than relying on patents). China offers these firms the chance to exploit additional advantages from low-cost production and economies of scale. These firms are therefore the best placed to generate innovations. The increasingly sophisticated R&D centers of Western and Japanese multinationals will also play a role in fostering innovation, mainly by training technical staff who may eventually set up their own firms.

Notes

1. For a detailed description of China's position in the IT production value chain, see G. Gaulier, F. Lemoine, and D. Ünal-Kesenci, "China's emergence and the reorganization of trade flows in Asia," Paris: *Centre d'études prospectives et d'information internationals*, Working Paper 2006–05 (March 2006), http://www.cepii.fr/anglaisgraph/workpap/summaries/2006/wp06-05.htm.
2. Ministry of Commerce, "海关公布年度出口 200 强 华为居民营企业首位" [Huawei No. 1 in annual list of 200 leading exporters published by Customs], 16 June 2006, http://foreigntrade.mofcom.gov.cn/aarticle/c/200605/20060502198530.html. Note that China's failure to produce global technology brands may simply reflect its relatively early stage of development. In their early stages of development, South Korea and Japan put up high barriers to foreign investment, but their firms relied heavily on licensed technology from abroad. It is possible that the end result of these two strategies for technology adoption (liberal FDI regime versus licensing) could be the same.
3. This discussion focuses on civilian IT research. The military IT research system is significant but so segregated from the civilian sector that it requires separate treatment. For an introduction to Chinese military R&D, see B. Gill and J. Mulvenon, "Chinese Military-Related Think Tanks and Research Institutions," *China Quarterly*, 171 (2002), pp. 617–624. See also E.S. Medeiros et al., "A New Direction for China's Defense Industry," RAND,

2005, http://www.rand.org/pubs/monographs/2005/RAND_MG334.pdf. This important study argues (pp. 205–251) that a "digital triangle" has begun to emerge linking commercial IT firms, state-funded R&D institutes, and the military.

4. Ministry of Information Industry, "国务院关于印发鼓励软件产业和集成电路产业 发展若干政策的通知" [State Council Notice on Policies Encouraging the Development of the Software and Integrated Circuit Industries], State Council Document 18, 17 December 2005, http://www.mii.gov.cn/art/2005/12/17/ art_80_1663.html.
5. National Bureau of Statistics and MOST, *China Statistical Yearbook on Science and Technology 2006* (Beijing: China Statistical Press, 2006), p. 222 with author calculations. It is likely that this figure includes spending by firms making TVs, DVD players, and household appliances, which are considered "IT" firms in MII's official annual catalog of the 100 largest domestic IT companies. See "19届电子信息百强名单" [The Nineteenth Edition of Top 100 Companies of Electronic Industry (in China)], 16 December 2005, http://www.mii.gov.cn/art/2005/12/17/art_65_1665.html.
6. In 2005, only five authentic IT companies (as opposed to consumer-electronics firms) in China spent more than USD 100 million on R&D: Huawei (USD 579 million), ZTE (USD 239 million), Lenovo (USD 183 million), Founder (USD 155 million), and Shanghai Alcatel-Bell, a joint venture (USD 111 million). Author calculations based on MII list of 100 top IT companies, ibid.
7. Scott Kennedy (Indiana University), personal communication.
8. See e.g., A. Bhide, "Venturesome consumption, innovation and globalization," paper presented at the CESifo Venice Summer Institute, July 2006, http://www.cesifo-group.de/pls/portal/docs/PAGE/IFOCONTENT/BISHERIGESEITEN/CESIFO_INHALTE/EVENTS/SUMMER_INSTITUTE/VSI06/VSI06-PAPER-ECPERF/VSI06_EP_BHIDE_0.PDF.
9. Domestic vendors' share of the mobile telecoms equipment market: Norson (HK) Information Technology Ltd, "China GSM Industry Supply Chain and Competition Analysis 2006 Q1," July 6, 2006. Data for the fixed-line sector: Ted Bean (BDA China Limited), personal communication.
10. For a detailed and balanced journalistic account of Huawei and its success in international markets, see K. Kuo, "China's Telco Titan grows," *Red Herring*, 13 February 2006.
11. ZTE, "二零零五年年度报告正文" [Text of Annual Report for 2005], 7 April 2006, http://www.zte.com.cn/main/about/Investor%20Relation/report/index.shtml?catalogId=12075.
12. D.B. Fuller and E. Thun, "China's Global Path," *World Busines* (July–August 2006); and Fuller, "The New Bamboo Network," *China Economic Quarterly*, Q3 (September 2006), pp. 29–33.
13. China Semiconductor Industry Association (CSIA), "Analysis of China's Integrated Circuit Industry," China Semiconductor Industry Association via Xinhua News Agency, 15 February 2006.
14. Data for Intel and TSMC from company websites. Data for China from CSIA, "Analysis of China's Integrated Circuit Industry."

15. "Semiconductor Report," United States Information Technology Office (USITO), December 2003, see www.usito.org.
16. China Semiconductor Industry Association (CSIA), "Analysis of China's Integrated Circuit Industry."
17. Liu Jiren, founder and chairman of Neusoft, told the author that building the software infrastructure for state enterprises and government agencies could keep domestic software firms busy "for 20 to 30 years." A. Kroeber, "Now for the Soft Part: China Tries the India Thing," *China Economic Quarterly*, 2001 Q1 (March), pp. 34–40.
18. This figure most likely also includes revenues from Dalian-based call centers that serve Japanese customers. T. Miller, "Bangalore Wannabe," *China Economic Quarterly*, Q2 (June 2005), p. 34.
19. "Microsoft deal to boost domestic sector," *People's Daily Online*, 27 April 2006, http://english.peopledaily.com.cn/200604/27/eng20060427_261621.html.
20. Sumner Lemon, "WAPI supporters ready a last stand in China," IDG New Service, 8 March 2006.
21. For a more detailed overview of China and its standards policies, see the following three papers from the National Bureau of Asian Research: R.P. Suttmeier et al., "Standards of Power? Technology, Institutions and Politics in the Development of China's National Standards Policy," NBR Special Reports, June 2006, http://www.nbr.org/publications/issue.aspx?ID=163; S. Kennedy, "The Political Economy of Standards Coalitions: Explaining China's Involvement in High-Tech Standards Wars," *Asia Policy*, 2 (July 2006), pp. 41–62, http://nbr.org/publications/asia_policy/AP2/AP2_Kennedy.pdf; and C.Y. Zhao and J.M. Graham, "The PRC's Evolving Standards System: Institutions and Strategy," *Asia Policy*, 2 (July 2006), pp. 63–87, http://nbr.org/publications/asia_policy/AP2/AP2_Zhao.pdf.
22. See J. Lipes, "China Mobile pushes for W-CDMA standard over home-grown TD-SCDMA," *Xinhua Financial News*, 5 December 2006, http://www.forbes.com/markets/feeds/afx/2006/12/05/afx3231227.html.

3

Nanotechnology Research in China

Chunli Bai and Chen Wang

3.1 Introduction

Research activities in the field of nanoscience and nanotechnology have flourished in China in the past decade. Due to the encouraging achievements made by Chinese researchers in both basic research and industrial applications, there is an increasing awareness in China of the potentials of nanotechnology. In this chapter we will provide an overview of the efforts being made in this country to develop research in nanotechnology.

China is one of the few countries that has been seriously engaged in nanoscience research activities since as early as the 1980s. Dating back from the 1980s to the present, Chinese scientists have explored nanoscience and nanotechnology in areas ranging from nanomaterials, nanodevices, and nanobiology to nano-characterization and nano-fabrication. Since 2001 nanotechnology has been prioritized as a strategic research field by the Chinese government. The Ministry of Science & Technology (MOST), the National Natural Science Foundation (NNSF), the Chinese Academy of Sciences (CAS), and the Ministry of Education (MOE) have been, and continue to be, the principal funding government agencies.

China has made a number of breakthroughs in studies on nanomaterials and related basic research. The most representative ones involve one-dimensional materials such as carbon nanotubes, nanometals, and single molecule detections. China's results in basic nanoresearch have earned recognition from the international scientific community, as will be discussed later in this chapter. As of the late 1990s, applied nanoresearch entered a stage of rapid development and has expanded to include societal impact and industrialization. However, though research on nanoscale devices has received a lot of attention in China, significant differences still exist between China and developed countries in the field of nanodevices and industrialization.

As an emerging multidisciplinary scientific field, nanoscience and nanotechnology provide a brand new platform of burgeoning possibilities and opportunities for technology innovations in many fields. The integration of basic research with applied research, coupled with exceptionally high expectations of the potentials of nanotechnology, has been a remarkable feature of the development in nanoscience and nanotechnology in China over the past two decades. Endeavors in this field have contributed to improve the traditionally weak atmosphere of interdisciplinary collaboration. This can be considered an additional benefit to the academic community at large.

The inception of nanoresearch in China can be traced back to the mid- and late 1980s when scanning probe microscopy (SPM) was introduced and developed in China as a novel technique for analyzing surface structures with nanometer scale resolution. The importance of the technique was soon recognized and generated broad interest in studying a wide range of materials at the nanometer scale. Researchers at CAS and different universities have developed several versions of SPM, while different research groups have demonstrated the fabrication of nanoscale features on surfaces using this technique. In addition, studies of a range of single molecules and molecular assemblies, with the aim of advancing the ultimate resolution power of material surfaces, have also been very encouraging. These exploratory studies have helped stimulate an awareness of the research into nanoscience and nanotechnology in China and the need to develop new techniques for revealing novel properties of nanostructures.

Material research dedicated to ultra-fine oxide particles was also approaching nanometer scale in the late 1980s at about the same time as the development of SPM, and novel properties associated with this scale were discovered. The direct applications of these pioneering activities led to the founding of a number of enterprises with products ranging from scanning probe microscopes to coating-related merchandise. Even though these manufactured products have not had significant commercial success, the experiences have provided hope for future entrepreneurship based on new intellectual properties.

There are numerous indicators of the advancement of nanoscience activities in China. Since 1990, dozens of international and national nanotechnology conferences have been organized. Scores of cooperation agreements have been signed between Chinese and foreign research institutions and enterprises. In recent years, the number of nano-related scientific articles by Chinese scholars has increased and Chinese scholars currently rank high in international statistics indicating the volume of published nanoscience papers. Patent applications and technological transfer are increasing. Among China's "Top 10 Science and Technology Advances in the News" of 2005, two were related to nanoscience and nanotechnology.[1]

Yet, in spite of all the progress made in the last two decades, China still has a long way to go in catching up with the leading nanotechnology nations. The country is still lacking in key technological innovations and has a limited infrastructure for nanoscience and nanotechnology research. Moreover, intellectual property protection as well as environmental and workplace safety have only recently begun to be prioritized.

3.2 National objectives of nanotechnology in China

In this new century, the Chinese government is committed to promoting scientific innovation in order to improve the quality of life of the Chinese people. The Ministry of Science and Technology focuses on both basic and applied research, with particular emphasis on the planning of certain strategically key areas. Nanotechnology is regarded as one such key research area that could have an impact on several disciplines.

"Guidance for National Development," issued by the Chinese government in 2001, designated nanotechnology as an R&D priority. Another important document, "Compendium of National Nanotechnology Development (2001–2010)" was jointly issued by MOST, MOE, CAS, NNSF, and the National Development and Reform Commission (NDRC) in the same year.[2] According to these documents, basic studies within nanoscience and nanotechnology are envisioned to continue while, at the same time, involvement from applied sectors is strongly encouraged. Preference should be given to national strategic tasks and market demands. High priority should also be given to ways of stimulating effective transfer from technological development to industrial applications.

In October 2000, MOST founded the National Steering Committee for Nanotechnology and charged it with undertaking the development of nanoscience and nanotechnology in China. At present, the committee consists of representatives from MOST, CAS, NNSF, MOE, NDRC, the Ministry of Finance, the Chinese Academy of Engineering, the Ministry of Liberation Army General Supply, the National Planning Committee, the National Economic and Trading Committee, the Science and Technology Committee and the National Defense Committee, as well as over 20 scientists from different institutions. The committee's mission is to act in a planning, coordination, and consultative capacity for nanotechnology projects in China at the national level. The "Medium- and Long-term S&T Development Plan," published by the State Council in early 2006, earmarked nanomaterials and nanodevices as an important goal for developing advanced materials.[3] Nanoscience is also listed as one of the four major research plans in basic research for this period of time. The following aspects of nano-related R&D in China have been chosen as focus areas:

1. Developing core technology of nanoelectronics and nanobiology
2. Creation of new functional materials and commercialization
3. From MEMS to NEMS, that is, reducing typical electro- or mechanical-device dimensions from micrometre to nanometre scale.

The directions of nanotechnology in the Eleventh Five-Year Plan (2006–10) focus on applications and devices in the biomedical area. In the first stage of research, early detection of diseases, low-cost medicines, and patient-friendly treatment facilities are listed as priorities. In the realm of device applications, projects on developing novel types of molecular-based sensors for electrical and optical devices have already been implemented. Polymeric composite materials with nanoscale designing precisions are also being extensively exploited for advanced materials for secondary battery, functional coating, energy-saving materials, and to supplement more costly materials.[4]

3.2.1 Government funding

In the late 1980s, NNSF started to fund research on instrumentation development such as scanning probe microscopy. The NNSF and MOST jointly established guidelines, policies, and blueprints for basic research in nanoscience. From 1990 to 2002, nearly 1000 projects were implemented for a broad range of nanotechnologies by national ministries, with funding of approximately USD 27 million. In addition, NNSF awarded approximately USD 12 million to nearly 1000 projects in the same time period.[5]

The total financial support for nanotechnology and nanoscience research from the Chinese government for 2000–04 was approximately USD 230 million.[6] When this figure is assessed in comparison with the world's leading nanotechnology nations, one can surmise that the progress by Chinese nanoresearchers has been achieved with relatively modest funding (see Figure 3.1). In the United States, for example, the government spent an estimated USD 3 billion on nano-related research from 2001 to 2004. Japan's spending was estimated at USD 2.9 billion and the European Union's at USD 2.2 billion during the same period.[7]

When assessing nanotechnology expenditure worldwide, it is useful to bear in mind that even the developed countries have only recently started to make massive investments in nanotechnology R&D. The three biggest investors, the United States, Japan, and the European Union were investing less than USD 150 million each in 1997 and the invested sums doubled twice between 1997 and 2005.[8]

Figure 3.1 Distribution of government support among subfields of nanoscience and nanotechnology in China 2001–05.

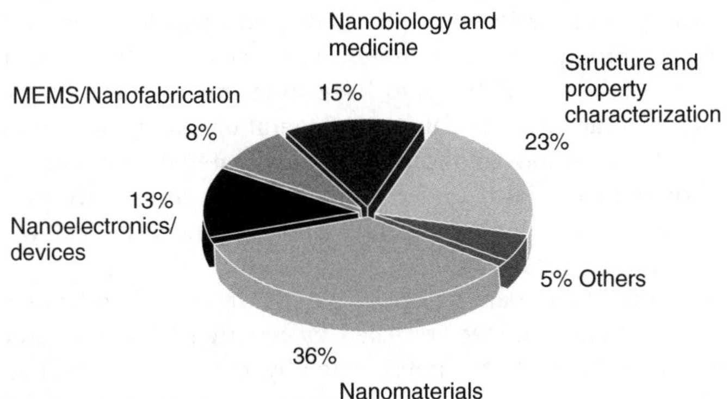

Source: Longxin Ou and Hongwei Dong, National Center for Nanoscience and Technology, China.

From 2006 to 2010, NNSF will further strengthen its support for China's nanoresearch, together with support from MOST and CAS, concentrating financial and human resources on tackling some of the key fundamental problems of nanoscience and promoting international collaboration and exchange. The application-oriented projects will also continue to be supported by MOST and CAS. In addition, several hundred enterprises are engaged in R&D of nano-related products, ranging from various nanoparticles to coating materials and detection instruments.

3.2.2 Safety issues and standards

Alongside the growing expectations in China of nanotechnology's potential in contributing to technological improvements, concerns have been expressed about the potential hazard of nanomaterials to the environment and human health. Early in 2004, a national special committee was created for laboratory accreditation under the auspices of the China National Board for Laboratories. This work is aimed at strengthening the metrological capabilities of the research facilities in public institutions as well as manufacturing sectors that are engaged in nanotechnologies.

To address concerns relating to nanotechnology and health, MOST, NNSF, and CAS have supported projects involving collaboration between materials scientists and experts on epidemic diseases. A national basic research project on the biological effects of artificially produced nanostructures was initiated in 2006. Addressing these concerns quantitatively will require dedicated scientific investigation, which may prompt new and interesting research directions. The preliminary results on safety issues have so far mainly been based on research conducted under laboratory conditions. Systematic evaluations of the hazardous implications of nanomaterials and nanostructures for human health under ordinary living conditions are still being carried out, both in China and elsewhere in the world. Therefore, it is too early to draw conclusions.

Due to the preliminary nature of the findings based on laboratory tests, the potential impact revealed by reported laboratory studies should be explained to the public carefully, to avoid false assurances or unnecessary alarm. Many more rigorous investigations are definitely needed. They are important because they could also set a good example of responsible activities by the research community when exploring cutting-edge technologies, which could ultimately benefit society. In addition, they are significant because they could serve to dispel suspicions abroad that China might be gaining an advantage in nanotechnology research due to lax ethical or safety standards.

Increased attention is being paid in China to workplace safety in conjunction with nanomaterials production. The relevant technical standards that may lead to future regulative actions are being studied. This involves a joint effort in examining the existing regulations of commercial products from a new perspective and preparation of the necessary reforms. These examinations cover textiles, health products, construction coatings, and so on. A number of major projects focusing on environmental and workplace safety issues in nanoproduction are currently supported by MOST, NNSF, and CAS. Moreover, members of the academic community have been called upon to join forces with government agencies to provide a genuine assessment of the social impact of the emerging technologies. This will provide guidance for both academia and industries.

Efforts directed toward standardization in China led to the publication of eight national standards in 2004, which were adopted in April 2005. The published standards consist of a glossary, four standards for nanoparticle products (nickel powder, zinc oxide, titanium dioxide, and calcium carbonate), and three for testing (gas absorption by BET method, photon correlation spectroscopy, and the granularity of nanopowder by XRD method). Moreover, four other standards (general rules for nanometer scale length measurement by SEM, sample preparation dispersing procedures for powders in liquids, and two standards on particle size analysis) were published in 2006. Over a dozen standards are currently being studied (see Figure 3.2).

Endeavors to set up measurement standards for nanostructures resulted in the initiation of the National Nanotechnology Standardization Committee in March 2005. This technical committee consists of experts in metrology, materials scientists, and standards administration officers, among others. Experimental work is already underway to set up a length standard for nanometer-scale objects. It has been realized that the typical time needed for developing standards is rather long (five years or more). Protocols that may lead to reliable

Figure 3.2 Nanostandards currently under study in China.

1. Measurement of shape, size, density, crystallization, chemical compositions, and structure of nanomaterials

- Expression methods of particle size distribution
- Expression methods of particle size classification
- Measurement of distribution of aperture of nanoporous
- Measurement of surface morphology of nanomaterials
- Static image analysis of particle size
- Distribution of interface atoms of nanomaterials tested by HRTEM test method
- Purity test methods of nanomaterials
- Composition and structure analysis methods of nano thin films

2. Testing methods of properties for nanomaterials

- Measurement method of contact angle
- Resistance and susceptibilities testing methods of nanomaterials
- Evaluation method for antibacterial inorganic nanomaterials
- Testing methods of adsorption structure for phthalocyanine and porphyrin molecules on the solid surface
- Testing method of the character of super-hydrophotic for nanomaterials
- Measurement of photolysis efficiency for self-clean nanofilms

measurement techniques are being seriously discussed in order to speed up the standardization process. The issue of adapting existing standards and regulations to newly evolved nanotechnologies has been under review in the process. The endeavor covers a wide range of disciplines, including construction coating, health care products, and textiles.

In recent years, there have been public announcements that some so-called nano-products have been banned from the market due to failure to comply with current regulations, such as misleadingly exaggerating the products' benefits. It is worthwhile noting that the lack of metrological standards for nanometer-scale structures is hindering attempts to update the currently enforced regulations on many so-called nano-products on the market.

There is also a growing awareness of the issue of intellectual property (IP) among nanotechnology researchers. The Chinese nanoresearch community faces similar problems to researchers within other disciplines in China regarding IP, the protection of IP and the implementation of regulations intended to protect IPR.

3.3 Nanoresearch institutions and researchers

More than 20 institutes of CAS, 50 university institutes, and several hundred commercial enterprises, involving a total of more than 3000 researchers, are working in the fields of nanoscience and nanotechnology in China. The dominance of CAS is overwhelming;[i] for example, in 2004, CAS institutes published more nano-related papers (1533) than any other institution in the world.[9]

Growing interest and support from industries, especially in the private sector, has been evident in recent years. For example, the Tsinghua-Foxcom Nanotechnology Research Center (TFNRC) was launched as an endowment from Taiwanese Foxcom Corporation in December 2003. TFNRC, led by Fan Shoushan, originally received funding of approximately USD 37 million. One of its main functions is to focus on technology transfer in nanoscience and nanotechnology.[10]

3.3.1 Major institutes

As a result of efforts to stimulate multidisciplinary research and development by building research and technology platforms, three national centers have been established. The National Center for Nanoscience and Technology (NCNST) was established in Beijing in 2003. The National Engineering Research Center for Nanotechnology (NERCN) was started in Shanghai around the same time and its construction is due to be completed in 2007. The China National Academy of Nanotechnology and Engineering in Tianjin (located on the premises of the Nanotechnology Industrialization Base of China, founded in 2000) was established in May 2005.

The NCNST, administered by CAS and headed by Bai Chunli, has been commissioned to focus on basic and applied research. Its mission is to be a public technology platform (for nanofabrication and characterization services), with the task of promoting new fronts in nanoscience, establishing databases, providing consultation for

[i] The high number of CAS publications is partly due to the unique nature of CAS; formally CAS is one entity administrating about 100 research institutes. These same statistics show three Chinese universities among the world's top ten institutions in nanopublications: Tsinghua University as fourth, University of Science and Technology of China—ninth, and Nanjing University—tenth, but these three together publish roughly only three quarters of the number of nanopapers published by CAS institutes.

government and industry, providing training for young scientists, promoting government–institution–industry links, and increasing opportunities for international collaboration.

The latter two centers are supported with funding from central and local governments and are equipped with state-of-the-art facilities for nanotechnology research. They serve as platforms for applied research and the commercialization of technologies from within China and overseas and the facilitation of international cooperation. The NERCN has received a total of USD 25 million of funding. One distinctive feature of NERCN is its mission to promote technology licensing from overseas.[11] The current research activities in the China National Academy of Nanotechnology and Engineering include studies on such aspects as developing metal and metal oxide thin films, MEMS technology and related applications. The establishment of these centers reflects the emphasis on the dual task of strengthening basic studies and stimulating transfer of knowledge to manufacturing domains.

Other important nanoresearch institutions are the major research institutes of CAS (e.g., Institute of Physics, Institute of Chemistry, and Institute of Metal Research in Shenyang) and a number of universities such as Beijing University, Tsinghua University, and Xiamen University.

3.3.2 Nanoresearchers and nanoeducation

More than 20 centers dedicated to nanoscience and nanotechnology have been established in universities and research institutes with funding from the state and local governments during the last two decades. Over 90 percent of the human resources involved in nanomaterials and nanotechnology research are located in the institutes under CAS and universities.[12] These institutions have trained a large number of young scientists and research teams. Various topics relating to nanoscience and nanotechnology have been offered as courses to undergraduate and graduate students in universities, as well as numerous short courses and seminars. Many technical books relating to nanoscience are available in Chinese for general readers such as college students.

The growing number of PhD dissertations relating to nanoscience and nanotechnology is a clear indicator of the progress in the education field. For example, PhD degrees in nanotechnology are offered as a multidisciplinary program in Beijing University. Students can join the program from the existing PhD programs in Physical Electronics, Physical Chemistry, and Condensed-Matter Physics. University-sponsored nanoresearch institutes have also offered graduate programs in related fields. A case in point is the Research Institute of Micro/Nanoscience and Nanotechnology of Shanghai Jiaotong University, which offers graduate courses in nanoscience and nanotechnology in its Master of Engineering program. University courses relating to nanoscience and nanotechnology have been extended from the current curriculum on quantum physics, material sciences, physical chemistry, microelectronics, and so on.

There seems to be a general consensus among educators that the multidisciplinary nature of nanoscience makes it more suitable for students who have already received high-level graduate training than for undergraduates, who are just beginning their formal scientific education. These graduate students will provide the main thrust for China's nanoresearch in the coming years.

3.3.3 Publications

In recent years the number of scientific nano-related articles by Chinese scholars has increased dramatically at an annual rate of 30–40 percent. Chinese scholars rank high in international statistics indicating the volume of published nanoscience papers.[ii] Based on a study examining research results published with a Chinese address in three core international nanotechnology journals, China ranked second after the United States in 2004 and 2005.[13] Statistics of the Scientific Citation Index also show China as the second largest single country after the United States in publication of nanotechnology articles.[14] Articles by Chinese nanoresearchers have on occasion also been published in the world's leading scientific journals such as *Science* and *Nature*. While in 2004 China's world share of all scientific publications was on average 6.52 percent, its world share in nano-related publications was 8.34 percent.[15] However, the same study also shows that the articles of the Chinese nanoresearchers are less frequently cited than the ones written in, for example, English-language

[ii] Researchers who have analyzed the rise in nanotechnology articles by Chinese scholars point out that the definition of a nanotechnology article is problematic because of the interdisciplinary nature of the subject. See for example P. Zhou and L. Leydesdorff, "The Emergence of China as a Leading Nation in Science," *Research Policy*, 35 (2006), pp. 84–85.

countries or in countries that use the Latin-based alphabet. But it would appear that this is also changing and the citation rate of Chinese publications is growing.[16]

However, nanotechnology expertise in China is still concentrated in two dozen or so universities and research institutes. Though there are over 1000 Chinese institutions contributing to international and domestic publications, the majority of published works come from authors in twenty or so institutions. Another distinct feature of Chinese nanopublications is that input from the industrial sector is missing. This situation is expected to improve as industries allocate more resources to nanotechnology R&D.

3.3.4 International collaboration

The development of nanoscience and nanotechnology has benefited greatly from academic exchanges between China and scientific communities abroad. In the past 15 years, dozens of international and national conferences, covering a wide range of topics in nano-related fields, have been organized in China. Nearly 1000 participants from international and domestic institutions attended ChinaNano2005 in Beijing in June of that year.

The strengthened international collaboration is evident from the establishment of joint research projects, research groups, and bilateral symposiums. For example, as of 2006, the German corporation BASF[iii] had 51 scientific projects underway in China with 35 universities and research institutes, including macromolecule material science, organic compounds, industrial catalyst research, nanotechnology, and biotechnology.[17] International collaboration has also been taking place between, for instance, NCNST and the Mitsubishi Corporation of Japan and Merck of Germany to develop composite materials using nanoparticles. Another illuminating example is a joint research center, Sin-China Nanotechnology Center, that was founded in 2005 in the Beijing University of Chemical Technology by NanoMaterials Technology (NMT) Pte Ltd of Singapore. The center is broadening and deepening the research on high-gravity technology in collaboration with renowned international companies including BASF, 3M, and Weyerhaeuser.

[iii] Major R&D initiatives of BASF are linked to the Sino-German Research and Development Fund.

The collaboration between international and Chinese research groups is encouraging, setting excellent examples for effective international collaboration on academic exchanges. However, one should note that international collaboration mostly takes place on an individual basis. At present there are still only a limited number of joint projects in nanoscience actually being implemented; many of them are in the form of exchange of graduate students and scholars. Though official frameworks for international collaboration have been set up in many areas and by many agencies, there is still much to be done to stimulate meaningful international collaboration, for example, by improving the research infrastructure in China and the research quality of Chinese scholars. China is a developing country with rather limited resources and capacity for developing extensive international collaboration. In addition, full-fledged interaction between China and the outside world has only been taking place in the field of science and technology for a couple of decades. Genuine scientific collaboration requires trust and appreciation that can only be built up over time. It will also require patience, frankness, and tolerance from all parties to bridge different cultural, social, and political factors. Nevertheless, international collaboration has definitely made great strides in the past decade or so in the fields of nanoscience and nanotechnology, and there is every reason to be optimistic that such progress will continue with increasing communication, mutual understanding, and recognition.

3.4 Nanoresearch in China

The R&D activities involving nanosciences in China can be categorized into materials, devices, bio-effects, and measurement techniques. Material-related research, dominated by the major universities and research institutes of CAS, is the most advanced in China, also by international standards. After the concept of nanoscience and nanotechnology was introduced in China, research focused on nanomaterials for a decade or so. In this century, the emphasis has been shifting to research and development of nano-sized devices, electronics, and biology, in line with the guidelines of recent national research plans. This has also been the trend in other leading nanoresearch countries.

The advances in synthesizing a wide range of low-dimensional structures, carbon nanotubes and other nanowires, nanoparticles, and composite materials best reflect the successes of Chinese researchers within nanomaterials. Some progress has already been achieved in industrial-scale productions and these products are becoming commercially viable.

3.4.1 Nanomaterials research

Studies on carbon nanotubes are representative of progress in the fast evolving field of low-dimensional material preparations. These all-carbon tubes are just a few nanometres in diameter, which makes them comparable in girth to deoxyribonucleic acid (DNA) molecules, and come in either single- or multi-walled varieties with a nesting of carbon shells resembling the structure of a retractable antenna. A number of Chinese groups working on carbon nanotubes are clearly in leading positions internationally. Back in 1996, a research group led by Xie Sishen at the Institute of Physics of CAS invented a template growth method, by which the growth direction could be controlled in the preparation of a directional carbon nanotube.[18] Groups from the Institute of Physics, Beijing University, and Hong Kong University of Science and Technology also actively engaged in the pursuit of the record-setting work of the longest and thinnest carbon nanotubes. In 2002, a group led by Fan Shoushan in Tsinghua University made yarns of carbon nanotubes.[19] After appropriate heat treatment, these pure carbon nanotube yarns are eventually able to be woven into a variety of macroscopic objects for different applications, such as bullet-proof vests and materials that block electromagnetic waves.[20]

The production of a large quantity of single- and multi-walled carbon nanotubes has been achieved in several laboratories, including the Institute of Physics, Tsinghua University, and the Institute of Metal Research of CAS. These efforts are envisioned as an important step toward industrial application in areas such as functional composite materials. With further advances in studies on nanotubes regarding issues such as chirality and length homogeneity, applications are expected to be expanded to broader interests such as microelectronic and optical devices.

A wide range of nanoparticles has been explored. This is by far the most advanced field in China both in basic research and industrial-scale productions. Many reports have demonstrated that the morphologies and compositions of the nanoparticle can be controlled with high precision. In keeping with the variety of technological applications, the basic studies focus mainly on the understanding of thermodynamics and kinetics for preparing nanostructures, and the correlation between structural characteristics and novel properties. These fundamental issues are critical for designing novel nanostructures, and more importantly, for converting the knowledge of nanoscience to any practical applications. In addition, functionalization of surfaces with nanoscale structures, leading to improved properties such as hydrophilicity or hydrophobicity, has also been demonstrated. Applications of newly developed materials and techniques resulting from nanoscience to mainstream industries are eagerly anticipated, as well as being seriously debated, in the hope of generating more efficient and sustainable businesses.

Within the category of metallic materials, a group in the Institute of Metal Research of CAS led by Lu Ke discovered the superplastic property of nanostructured copper. Copper that has been nanotechnically recast can be elongated at room temperature to more than 50 times its original length without breaking. Subsequent studies further showed that so-called copper growth twins, another kind of nanocopper consisting of a specific type of crystalline microstructure, have a tensile strength about 10 times higher than that of conventional coarse-grained copper, while retaining an electrical conductivity comparable to that of pure copper.[21] In another study by this group, a greatly enhanced diffusivity of chromium was identified in the surface mechanical attrition treatment (SMAT) of Fe at 350°C, which is about 300–400°C lower than the conventional treatment temperature.[22]

3.4.1.1 Carbon nanotubes

As touched upon previously, the studies on preparation and characterization of carbon nanotubes can be viewed as an example of the endeavors toward nanoscale. Being the first discovered linear structure with a diameter smaller than one nanometer and length as long as tens of millimeters, carbon nanotubes are expected to possess many enhanced properties unseen in traditional materials, such as high mechanical strength with greatly reduced weight as well as high thermal and electrical conductivity. Carbon nanotubes have been extensively researched for application in areas such as nanodevices (sensors, logic circuits, etc.), opto-electrical materials, and so on. The first report on controlled growth of carbon nanotubes by Xie Sishen of the Institute of Physics of CAS in 1996 was the result of several years' research on this novel material. This work immediately generated broad interest from colleagues in the fields of physics, chemistry, and materials science. Support from CAS, NNSF and MOST also facilitated these advances. For almost a decade, the research yielded many remarkable achievements, partly mentioned above. Moreover, one should note that the work on carbon nanotubes also contributed significantly to the much broader explorations of various nanostructures, including metal, semiconductor, and organic species. These activities played a major role in the basic research of nanomaterials, both in China and internationally.

The Synthesis and Characterization of SWNTs

Sishen Xie

Since their discovery, carbon nanotubes have been the subject of intensive research due to their unique structures and extraordinary properties. An ideal 1D material, a single-walled carbon nanotube (SWNT) can be constructed by rolling up a perfect graphene sheet to make a seamless cylindrical tubule with a diameter on the nanometer scale.

Figure 3.3 Large-scale SWNT non-woven material.

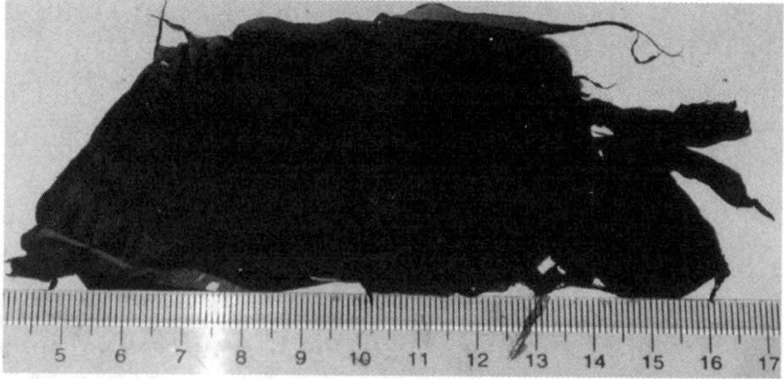

During the last decade, a great deal of effort has been exerted to studying the outstanding properties of SWNTs and realizing their potential applications in many fields, for example, as electron field emitters in panel displays, biosensors and carriers for drug delivery, enhanced materials in composites, candidates for the next generation of transistors, and so on. In realizing their novel applications, a great challenge lies in achieving the controllable preparation of SWNTs on a large scale. In my group, we started researching large-scale controllable preparation of SWNTs by chemical vapor deposition (CVD) in 1994.

Recently, we have been focusing on developing a continuous CVD system (called floating catalyst CVD), in which SWNTs are grown in a flowing gaseous feedstock mixture. In this way, we have successfully made SWNT nonwovens and rings, which provide further opportunities for the application of SWNT.

Single-walled carbon nanotube (SWNT) nonwoven

The SWNT nonwoven with an area of up to several tens of square centimeters resembles paper made of compact SWNT. Figure 3.3 illustrates a large-scale SWNT non-woven material with a compact structure. The SWNT nonwoven retains most of the outstanding properties of SWNT, such as high stability, high strength value, and electrical conductivity. The tensile tests on SWNT nonwoven reveal extremely high Young's Modulus of approximately. 0.4~0.7 TPa, which has potential applications in the mechanical reinforcement of composite materials.

Single-walled carbon nanotube (SWNT) rings

As a novel superstructure, SWNTs with annular geometry exhibits interesting transport properties when the electric current flows along the ring or a magnetic field passes through the hole in the ring. We improved the normal method of CVD and made significant advances in ether quantity and the quality of the rings. The typical morphology of the nanotube ring can be seen in Figures 3.4a and 3.4b, depicting the product with a high yield and perfect ring structure.

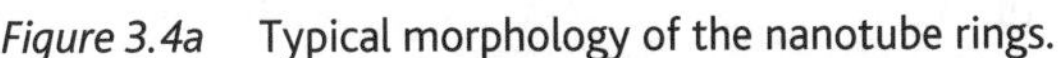

Figure 3.4a Typical morphology of the nanotube rings.

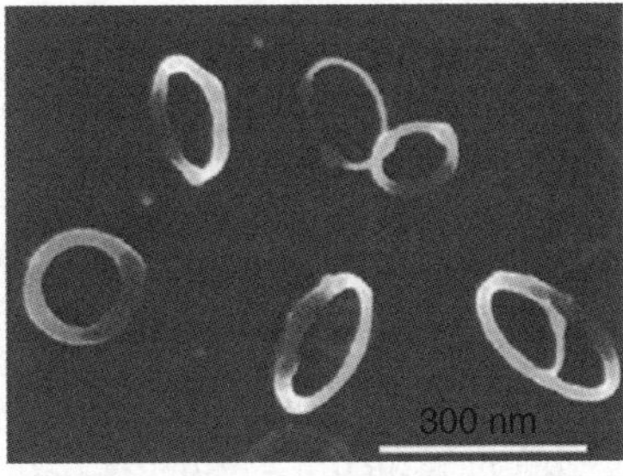

Rings typically have an average diameter of 120 nm and a narrow thickness distribution in the 15–30 nm range. The smaller diameter of the rings may have more interesting quantum effects and electronic properties. It would appear that the ring with the annular geometry is composed of SWNT bundles, decorated with small catalyst particles. The thickness of the ring is 25 nm. The SWNTs with a diameter of 1 to 2 nm are found to be in tightly packed hexagonal formation.

Offering a remarkable advantage for potential applications such as electronic devices, the SWNT rings could be deposited at a very low temperature on different substrates. This low temperature deposition allows the integration of the nanotubes with silicon and MEMS technology.

Professor Xie of the Institute of Physics of CAS, is an Academician of the Chinese Academy of Sciences.

Figure 3.4b Typical morphology of the nanotube rings.

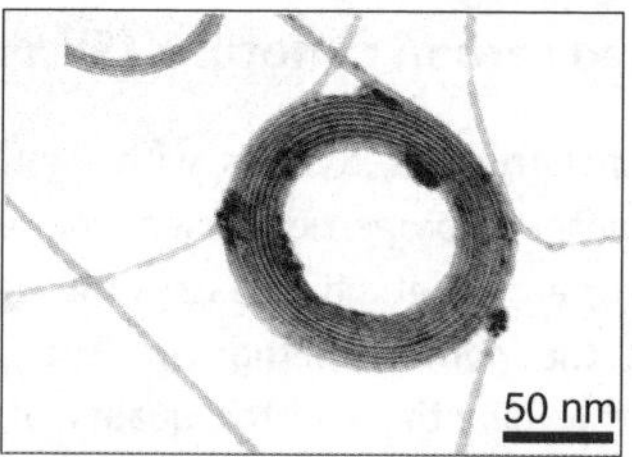

3.4.2 Nanodevices

While research on nanoscale devices has received a lot of attention in China, research has progressed relatively slowly because of limited research conditions, especially the availability of fabrication facilities at micrometer and nanometer scale. Although the existing fabrication facilities (the so-called top-down approach) provide support for researchers working on nanoscale devices, new concepts are definitely needed to support a broader base of users with non-standard research objectives. Such objectives stem from the novel properties (electrical, optical, and mechanical) discovered in nanostructured materials and the possibilities they offer for conceptually new products. Building macroscopically sensible structures utilizing nanostructured building blocks represents a different direction, a "bottom-up" approach. Communication between and unification of these two approaches constitutes a weak link in the chain of development from basic research to products in China, as is the case in many other countries.

The applications of nanotechnology in microelectronics and opto-electronics are clearly important. Coordinated efforts have been made to organize resources to tackle key techniques for nanoscale devices. The increase in research using some of the newest nano-structured materials looks set to lead to novel prototype electronic and opto-electronic devices. As part of the push to strengthen research capability in this area, the construction of clean-room device fabrication facilities in a number of nano-related research centers, for example the Nanocenter in Tsinghua University is underway. At the same time, sharing fabrication facilities among institutions has also proved to be a solution which is actively promoted through NCNST.

One widely pursued topic in nanodevice studies is related to the exploration of using molecular architectures as units for building up complex devices, such as molecular electrical conductive wires, transistors, electrical or optical switches, high-density data storage mediums, and so on. Several groups at CAS have made progress in the past few years, such as the single molecule studies led by Hou Jiangou at the University of Science and Technology of China (USTC) in Hefei, Anhui, and the high-density storage studies led by Gao Hongjun at the Institute of Physics of CAS.

3.4.3 Bio-effects

The biological effects of nanomaterials have, by and large, been examined by a number of groups across China. For instance, a group led by Shi Jianlin at the Shanghai Institute of Ceramics of CAS has demonstrated that nanoparticles with a very uniform diameter of only a few tens or hundreds of nanometres can carry and transport drugs in human blood vessels under the guidance of external magnetic fields. The drugs could also be released in a controlled manner at the designated locations, increasing the efficiency of the medicine and reducing the otherwise possible side effects.[23] In another study by Hu Jun et al. at Shanghai Jiaotong University, it was found that favorable binding of gold nanoparticles to single-stranded DNA decreases the likelihood of mispairing formation during annealing and clearly increases the specificity and yield of polymerase chain reaction

(PCR), one of the most important standard methods of molecular biology. Recent work carried out by groups in Yunnan University and Jilin University using nanocrystals of gold and magnetic materials modified with antibodies for detecting hepatitis B and HIV viruses has already gained clinical support and is being developed for commercial manufacturing.[24]

In particular, toxicology studies on nanomaterials have been carried out at the Institute of High Energy Physics and the Institute of Biophysics of CAS, as well as at the Academy of Military Medical Sciences. It should be noted that current activities are still limited to a small number of species such as fullerenes and nanoparticles. Substantially expanded research activities are needed in the future to address safety issues of common concern. Such studies are also viewed as playing a vital role in developing drug delivery agents using nanostructured materials. Efforts toward targeted drug delivery have been made in groups at the Shanghai Institute of Materia Medica and the Institute of Biophysics, among others.

One frequently raised subject concerning biological and medical-oriented studies is that biologists and medical specialists should be more seriously involved in order to effectively guide the research activities originating from other disciplines. This kind of collaboration could stimulate advances in both nanoscience and medical and biological sciences.

3.4.4 Measurement techniques

The improvement of measurement techniques is a vital part of nanoscience, as exemplified by the development of scanning probe microscopy, achieved with the support of CAS and NNSF. Over the years, the capabilities of resolving atomic lattices and molecular fine structures have been demonstrated by many research groups in China, such as research teams at the Institute of Chemistry and the Institute of Physics of CAS, as well as the University of Science and Technology of China. Research into using scanning probe microscopy as a metrology method at the nanometer scale has also been actively pursued by groups at the Chinese Academy of Metrology and USTC. In the meantime, groups at the Institute of High Energy Physics are conducting investigations into other detection techniques, such as using the strong light source of synchrotron radiation to detect cellular structures. A number of groups are engaged in studies to improve the resolution of fluorescence microscopy and other types of optical microscopes. It is hoped that this research may lead to new measurement standards that are of paramount importance for nanotechnology.[25]

3.5 Technology transfer and commercialization

Several hundred Chinese enterprises are engaged in nanotechnology. According to statistics collected from conferences and publicly announced projects, more than 500 enterprises are associated with nanotechnologies to some extent. These enterprises include vendors of nanopowders, coating materials, fabrics, and health products; many of them are small and medium-sized private companies.

Reliable information on the number of genuinely noteworthy nanotech companies in China is hard to come by. Arriving at an exact definition of a nanotechnology company is already problematic. The report "China Industrial Nanotechnology 2006" is nevertheless illuminating; it derives from an analysis of over 600 nanotechnology Chinese firms on the basis of their degree of engagement in development or production of nanotechnologies, their capacity to produce at the industrial scale and their attainment of ISO certification. As a result, 80 companies with an annual revenue or registered capital of over RMB 10 million (USD 1.25 million) were chosen to be included in the report as enterprises for foreign investors to keep an eye on for potential investment or licensing opportunities.[26]

3.5.1 Enterprises related to nanotechnology

As described previously, three national nano-centers have been established to promote the links between governmental institutions and industry. The National Center for Nanoscience and Technology in Beijing is dedicated to basic and applied research, while the National Engineering Research Center for Nanotechnology in Shanghai and China National Academy of Nanotechnology and Engineering in Tianjin focus on the industrialization of nanotechnology. For the latter two, collaboration with industries plays a vital role.

Most of the more important companies engaged in nanotechnology in China concentrate on nanomaterials. Several manufacturers,

Figure 3.5 The major areas for commercialization in nanotechnology.[27]

Nanomaterials	Nanodevices	Nanobiology
Composite materials	Sensors	DNA and protein chips
Multifunctional powders	Healthcare detectors	Chinese traditional medicine
	Storage and display	Tools for early diagnosis

such as Shenzhen Nanotech Port Co. Ltd, have achieved industrial production of carbon nanotubes at the scale of a few tens of tons per year of multi-walled carbon nanotubes. Motivated by continued efforts to search for low-cost and high-performance materials, the applications of carbon nanostructures, including nanotubes and fullerenes as additives in a variety of organic composite materials, have led to many interesting findings such as the improvement of mechanical, thermal, and electrical properties. Many of these kinds of studies are being actively pursued for industrial applications, including high-performance engineering polymers that can be used in automotives and oil-free lubrication materials for use in many mechanical devices.

Many of the most significant nanontechnology-related enterprises in China were founded with technology transferred from research institutions, and these enterprises have duly maintained relatively strong ties with the research community. Exploration of industrial applications has been pursued by a number of groups, as is evident from the following examples: Chen Yunfa of the Institute of Process Engineering of CAS, in collaboration with several domestic joint ventures, used nanoparticles to develop new construction coating materials and antibacterial materials. A group led by Chen Jianfeng at the Beijing University of Chemical Technology successfully commercialized the production technology of nanoparticles of $CaCO_3$ with particles sizes of 15nm–30nm by means of a high-gravity platform, with low-cost and high-volume production.[28] At present, nano calcium carbonate production lines using high-gravity technology have been put into operation for the production of construction materials in five enterprises in China, varying in capacity from 3000 to 10,000 t/a per line. In addition, a 1000 t/a Mg $(OH)_2$ production line was set up in Tianjin, while a 40 t/a commercial production line of cefuroxime axetil nanodrug powder was put into operation by the North China Pharmaceutical Group Corporation in 2004.

The successful transfer of technology, as exemplified above, has helped generate continuing interest in the commercialization of nanotechnology (see Figure 3.5). It is noteworthy that major industries and large enterprises have yet to play significant roles in exploiting nanotechnology, although awareness and expectations have greatly increased in recent years, and more substantial involvement in the form of contract research and joint projects is expected to accelerate in the future.

When assessing nanotechnology companies, it is useful to bear in mind that there is a popular misconception that nanotechnology is an industry or a sector—it is not. Nanotechnology is a set of tools and processes for manipulating matter that can be applied to innumerable kinds of manufactured goods in several industries. It is misleading to speak of a nanotechnology market, and it would be more useful to focus on how nanotechnology is being used across industry value chains, from basic materials to intermediate products to final goods. For example, sales of basic nanomaterials like carbon nanotubes are not expected to be significant globally in the coming decades. Nanotechnology's economic impact will emerge from how innovative nanotechnology processes are used, not from sales of the materials themselves. This is the challenge for China and for all nanotechnology nations.

3.5.2 Patents

The encouraging efforts on the part of researchers into nanotechnologies have resulted in a significant increase in applied and approved patents. The total number of approved nanotechnology patents in China increased from just over one thousand in 2002 to more than two thousand in 2004. Statistics indicate that by August 2006 a total of more than six thousand published patents relating to nanotechnology had been approved in China.[29] The majority of the nanotechnology patents have originated from research institutions and universities, while industrial companies file only less than one-third of the approved patents. However, Chinese researchers have been slow to file international patents, which is mostly due to a lack of financial support and awareness from the research bodies.

The number of patents by Chinese researchers issued in the United States is not significant; for example, in 2003 China just made it into the group of the top 20 nanopatent-producing countries in the U.S. Patent Office.[30] So there is room for great growth in the future.

Health-related nanotechnology research appears to be a notable exception in China's otherwise slow pace in filing international patents. With its 20 percent share of health-related nanopatents, China ranked second in the world in 2005, between the United States (33 percent) and Germany (13 percent).[31]

3.6 Conclusion: Analysis of level of nanoresearch in China

The strong performance of basic research into nanoscience and nanotechnology is continuing in China and the speed of knowledge transfer to practical applications is dramatically increasing. As described above, the most far-reaching and internationally acclaimed nanoresearch in China has been conducted in the fields of nanomaterials and some areas of characterization techniques, especially with regard to the development of low-dimensional nanostructures. This is partly the result of continued financial support for basic nanotechnology research for nearly two decades. In addition, an increasing number of university faculty members and students are engaging in research projects, joining the endeavors of research institutions. These changes in financial support and the research population are not restricted to the fields of nanoscience and nanotechnology. The Chinese scientific community, in general, is benefiting from similar improvements.

Up until now, research results in the areas of nano-electronics and nano-bio have been modest in China. The next stage of fast growth areas in nanoscience can be predicted to relate to biological effects, and various artificially fabricated structures and devices.

A substantial proportion of government support is used for improving or setting up research facilities that have become obsolete or have simply not been available. This situation is common among research communities worldwide at the onset of nano-related research activities. With the exploration of nanotechnology beyond the realm of nanomaterials into areas such as nanodevices, energy, and the environment, research capabilities have to be strengthened with both much-needed expertise and adequate facilities. The development of nanotechnology at the current stage is mainly embedded within the existing industries. There are expectations that nanotechnology could stimulate technological innovations for both domestic and international industries.

In the meantime, the involvement of major industries together with the growth of current nanotechnology enterprises will further shape the directions of nanotechnology in China. More administrative efforts are needed to coordinate research activities and increase the efficiency of technology transfer. Safety awareness and attention given to preparatory protection within nano-product development could serve as a new development model.

3.6.1 Future prospects: major areas of growth

It is clear that nanotechnology as a whole is still in its infancy in terms of generating economic impact. Publishing papers on nanomaterials, for example, does not yet mean that the properties of these materials have been fully examined, nor that there is knowledge about which applications could benefit from them. The next phase of research activities in China is expected to seek key technological breakthroughs mostly pertinent to industries in China, from simple value-added products to more sophisticated technological processes. In the long run, materials-based products will lead the way to a commercialization process, followed by medical applications and novel devices.

China has achieved an important role in nanotechnology research because nanotech is a so-called new area of research and China started doing nanoresearch around about the same time as the rest of the world. However, appreciable differences in the overall level of nanoresearch still exist between China and developed economies, especially in the area of nanoscale devices and the process of industrialization. China still has a long way to go to improve the overall competitiveness of its nanoscience and nanotechnology enterprises. Chinese scientists will continue to contribute to the quest for nanotechnology in many ways, both complementing as well as competing with the global community to advance our knowledge in this science. The growing involvement from enterprises will be significant for the advancement of nanotechnologies in China.

In the area of applied sciences, China has and will continue to have its own priorities based on national and societal needs, including increasing energy efficiency and developing low-cost drugs and new treatment methods for fatal diseases. There will be increased collaboration with international colleagues as well as international industries in the field of applied research. Such progress should be seen as a potentially important contribution to the sustained development and improvement of the quality of life of the Chinese people, which will also benefit the international community.

The nanoresearch communities in China and in the rest of the world are confronting similar challenges in this field. Efforts made in basic research are complementary on the global scale. The Chinese nanoresearch community, with very limited financial support compared to the developed economies, has already contributed appreciably to basic research. Some achievements are considered important breakthroughs in their respective areas, as exemplified by nanomaterials. More research on applied science will be encouraged in the future. With increased attention being paid to the field of nanoscience, continuous contributions from Chinese researchers can be expected, with increasingly innovative findings.

Notes

1. C.L. Bai, "Nanotechnology in China," *TWAS Research Update* #2, Academy of Sciences for the Developing World, August 2006, http://www.twas.org/.
2. M. Berger, "Recent Industry Studies Say that China is Quickly Becoming a Leading Force in Nanotechnology," *NanoChina*, 21 March 2006, http://www.nanochina.cn/english/index.php?option=content&task=view&id=589&Itemid=182.
3. State Council of the People's Republic of China, "国家中长期科学和技术发展规划纲要" [Outline of National Medium- and Long-term S&T Development Plan (2006–2020)], 9 February 2006, http://www.gov.cn/jrzg/2006-02/09/content_183787.htm.
4. Central Committee of Chinese Communist Party, "中华人民共和国国民经济和社会发展第十一个五年规划纲要" [Text of the Eleventh Five-Year Plan for the Development of the Economy and Society of the PRC], 16 March 2006, http://news.xinhuanet.com/misc/2006-03/16/content_4309517.htm.
5. C.L. Bai, "Ascent of Nanoscience in China," *Science*, 309 (2005), p. 62.
6. H.P. Jia, "Government raises nano-tech funding," *China Daily*, 10 June 2005.
7. European Union refers to EU-25 plus Switzerland. M.C. Roco, "International Perspective on Government Nanotechnology Funding in 2005," *Journal of Nanoparticle Research*, 7 (2005), pp. 709–710.
8. Ibid.

9. R.N. Kostoff et al., "The Structure and Infrastructure of the Global Nanotechnology Literature," *Journal of Nanoparticle Research*, 8 (2006), p. 306.
10. L.W. Liu, "Leading Nanotech Research Center in China," *Asia Pacific Nanotech Weekly*, article 23 (2004), http://www.nanoworld.jp/apnw/articles/2-23.php.
11. L.W. Liu, "China Setting up a Platform for Nanotechnology Commercialization and International Cooperation," *Asia Pacific Nanotech Weekly*, article 9 (2005), http://www.nanoworld.jp/apnw/articles/3-9.php.
12. H.C. Gu and J. Schulte, "Scientific Development and Industrial Application of Nanotechnology in China," in Schulte (ed.), *Nanotechnology: Global Strategies, Industry Trends and Applications* (John Wiley & Sons, 2005), p. 11.
13. P. Zhou and L. Leydesdorff, "The Emergence of China as a Leading Nation in Science," *Research Policy*, 35 (2006), pp. 93–94.
14. Data calculated from Scientific Citation Index website, http://scientific.thomson.com/products/sci/.
15. P. Zhou and L. Leydesdorff, "The Emergence of China," p. 100.
16. Ibid., pp. 88–89, 99.
17. *China Chemical Reporter*, "BASF Touts Nanotechnology in Shanghai Event," 26 May 2006, p. 8.
18. W.Z. Li et al., "Large-Scale Synthesis of Aligned Carbon Nanotubes," *Science*, 274 (1996), pp. 1701–1703.
19. K.L. Jiang, Q.Q. Li, and S.S. Fan, "Spinning Continuous Carbon Nanotube Yarns," *Nature*, 419 (2002), p. 801.
20. See C.L. Bai, "Ascent of Nanoscience," p. 62.
21. L. Lu, M.L. Sui, and K. Lu, "Superplastic Extensibility of Nanocrystalline Copper at Room Temperature," *Science*, 287 (2000), pp. 1463–1466.
22. W.P. Tong et al., "Nitriding Iron at Lower Temperatures," *Science*, 299 (2003), pp. 686–688.
23. Y.F. Zhu et al., "Stimuli-Responsive Controlled Drug Release from a Hollow Mesoporous Silica Sphere/Polyelectrolyte Multilayer Core-Shell Structure," *Angewandte Chemie International Edition*, 44 (2005), pp. 5083–5087. See also W. Zhao et al., "Fabrication of Uniform Magnetic Nanocomposite Spheres with a Magnetic Core/Mesoporous Silica Shell Structure," *Journal of the American Chemical Society*, 127 (2005), pp. 8916–8917.
24. L.Y. Wang et al., "一种乙肝病毒表面抗原和其单克隆抗体相互作用的 SERS 研究" [Surface-Enhanced Raman Spectroscopic Study on the Interaction of Hepatitis B Virus Surface Antigen and Its Monoclonal Antibody], 化学学报 [Acta Chimica Sinica], 60:12 (2002), pp. 2115–2119.
25. P.P. Zhu et al., "Computed Tomography Algorithm Based on Diffraction-Enhanced Imaging Setup," *Applied Physics Letters*, 87 (2005), p. 26.
26. Nabacus Blue Book Series, *China Industrial Nanotechnology Report*, sample, p. 2, http://www.nabacus.com/BlueBookCNsample.pdf.
27. S.S. Xie, "Fundings and networks for nanotechnology in China," presentation at the 2nd Workshop on Nanotechnology Networking and International Cooperation, "8th IUMRS Conference on Advanced Materials," Yokohama, 11–12 October 2003, p. 4, http://www.nims.go.jp/ws-nanonet/Speakers/Xie-ICAS-Chinar.pdf.

28. J.F. Chen et al., "Synthesis of Nanoparticles with Novel Technology: High-Gravity Reactive Precipitation," *Industrial & Engineering Chemistry Research*, 39:4 (2000), pp. 948–954.
29. Data calculated from State Intellectual Property Office website, www.sipo.gov.cn.
30. Z. Huang et al., "International Nanotechnology Development in 2003: Country, Institution, and Technology Field Analysis Based on USPTO Patent Database," *Journal of Nanoparticle Research* 6 (2004), p. 331.
31. D. Maclurcan, "Nanotechnology and Developing Countries. Part 2: What Realities?" *AzoNano Online Journal of Nanotechnology*, 19 October 2005, http://www.azonano.com/Details.asp?ArticleID=1429.

4

Energy Technology Research in China

Kejun Jiang[i]

4.1 Introduction

In the 1950s China established research institutes with Soviet expertise and aid. Fundamental technologies including energy supply and utilization technologies such as cutting machines, heating furnaces, small boilers, and blast furnaces were developed to support domestic economic growth. The first 125 MW power generator was made by Harbin Electric Motor Factory in the early 1960s. Since then, China has made significant progress in technology development, including development of basic energy utilization equipment and energy conversion equipment.

During the early years of economic reform, the focus of the government's science and technology (S&T) national programs was to develop ways to reduce the use of energy and conversion technologies, but over the years, high-efficiency coal-fired power generation, coal liquefaction, and boilers have received financial backing.

Owing to China's rapid economic growth, total primary energy consumption increased from 400 Mtoe (million tonnes of oil equivalent) in 1978 to 1560 Mtoe in 2005. By 2030 energy demand could reach 3 billion toe.[1] Coal has been and still is the major energy source, providing 70.7 percent of total primary energy consumption in 1978, and 68.9 percent in 2005.[2] Widespread energy shortages have been common in recent years. As a result, energy technology research and

[i] The author would like to thank Research Professor Hu Xuelian for her assistance during the process of writing this chapter.

development have become key components of overall technology research and development (R&D).

In the future, China's increasing energy demand will put enormous pressure both on the environment and the energy supply. China is predicted to become the world's largest emitter of carbon dioxide by 2010.[3] Moreover, the country's thousands of small- and medium-sized coal-fired plants operate with obsolete technology which is detrimental to the environment. Every month, three to four new coal-fired plants are built in China, all of which are adding significantly to the effects of pollution.[4] It has already become evident that the target for increased energy efficiency in the Eleventh Five-Year Plan of the Chinese government may be difficult to meet. In January 2007, Chinese officials announced that the 2006 target of 4 percent reduction in energy consumption per unit of GDP had not been reached; on the contrary energy consumption rose.[5] Therefore clean and high-efficiency energy conversion technologies, energy utilization technologies, as well as new and renewable energy utilization technologies are crucial for China. If new energy technologies can be produced at a reasonable cost, China's energy prospects, and indeed the prospects for the world as a whole, would be viewed in a different light. As a result of the "lock-in" effect[ii] of technology and investment, diffusion of new energy technologies should be more widely implemented at an early stage.

4.2 National objectives of energy-related research

In general, China's national technology development objectives focus on following international research in selected technology areas, and on reducing the gap between China and more developed countries. China is also making efforts to achieve breakthroughs in the areas where it has advantages to facilitate future economic development and national security. Clean coal technology is one such area. Energy has been targeted as a key aspect for the government's support.

China aims to safeguard its energy supply by developing both high-efficiency and alternative technologies. The country also needs to develop technology innovation that is directed toward its specific

[ii] "Lock-in" means that once infrastructure is adapted to suit certain technologies it is costly to alter the infrastructure.

needs, in addition to being internationally competitive. There is significant domestic demand in China for high technology in several energy-related fields. Areas of importance include coal-fired power-generation equipment, nuclear, and biomass utilization technologies, hydropower generation, as well as oil exploitation and coal-mining technologies. Of course, one must not forget the economic importance of the energy sector and energy technology manufacturing sectors, which are not only extensive but also high value-added industries, having an impact on employment.

The "Medium- and Long-term S&T Development Plan," published in early 2006 by the State Council, sets specific targets for energy technology development, which is listed first in the document. Within 15 years, Chinese researchers are expected to make breakthroughs in several energy-related fields. Examples listed in the document include energy conservation in industry, clean coal energy development and utilization, liquefaction, and polygeneration. In addition, the document targets oil and natural gas exploration and utilization in difficult, complex geographic areas, low-cost renewable energy development and utilization, ultra-large-scale transmission and distribution of electrical power, and grid security. Energy use for manufacturing major industry products is expected to reach or approach the level of advanced countries. Advanced energy technologies such as hydrogen technology and fuel cell, distributed energy supply technology, fast breeder reactor, and nuclear fusion were added to the plan in several rounds of expert discussions organized by the Ministry of Science and Technology (MOST).[6]

At the beginning of 2006, the State Council also announced its development plans in a document entitled "Several Suggestions for Accelerating Equipment Manufacture Development in China." In this paper, energy technology is again placed as the first focus technology and a specific aim is stipulated: By 2010, a group of large Chinese equipment manufacturing companies should be internationally competitive to meet the demands of energy, transport, and raw-material production. The energy equipment listed in the document includes large, clean, high-efficiency power generation systems, such as gigawatt (GW)-scale nuclear generators and several different technologies for clean coal and renewable energy. Equipment development for 1 MV super-high voltage and 800 kV direct current transmission system, mastering the manufacture of 500 kV and 750 kV transmission system technology, as well as large underground-integrated coal-mining technology and large offshore oil engineering equipment are included in the list of technology that need future development.[7]

4.2.1 Government steering

In 1983, the Large Equipment Leading Group under the State Council was established to promote advanced equipment manufacturing in China. The group decided to develop the 300 MW and 600 MW Qinshan Nuclear Power Plants, hydropower generators for the Three Gorges Dam, BaoSteel equipment, and Yanshan ethylene production equipment. A number of research institutes and industry manufacturers worked together and made significant progress on the technologies in question. Later, targets regarding technology R&D were incorporated into governmental research programs such as the 863 and 973 programs. Government R&D programs mainly focus on emerging and utilization technologies before commercialization. Technologies listed in these programs are predetermined by steering committees of experts in different fields.

For example, a steering group of specialists from research institutes, universities, and industrial enterprises selects the energy technologies that are to be included in the 863 program. The members of the group are selected through processes of expert and government nomination. For each round of new technology development planning, the steering group will have several rounds of meetings, as well as extended meetings that include more experts. Finally, the group will decide the technology list for the program. The government's organizing agency is MOST.

In addition to the expert groups, a new concept discovery research program was established under the 863 program to provide new theory and technology ideas for the whole program. This research program is under the administration of the National Natural Science Foundation (NNSF), with a two percent share of the total 863 program funding.[8] The technology selection process in the 973 program is similar to that of the 863 program; expert groups are organized for each of the research areas, including energy.[9]

4.2.2 Funding

Although the government is the main financer of energy R&D projects, specific funding earmarked for energy technology development is hard to identify. The leading agency in the government is MOST: It draws up the technology R&D plans and provides funding for the national programs such as the 863 program and 973 program. Energy technology-related R&D funding has always been a part of a national program or other research project funding.

Figures 4.1–4.4 list several sources of funding for energy technology research in China, together with the number of researchers involved. Figure 4.1 illustrates government funding for main research programs on energy technology, which has increased, with the exception of basic research, during the past few years. This is mainly due to an increase in the government budget and the huge demand for energy technologies in China. Among the main national programs, funding for the 863 program has risen rapidly. The energy technology research funded by this program focuses mainly on technologies that are nearing the commercialization stage, especially technologies already in use in other countries. These are also supported by industries with their own funding.

The main goal of the Torch plan is to spread new technology nationally, in other words to facilitate the use of new technology at the local level. Three of the plan's eight subsectors focus on energy technology. The total budget has increased owing to the growing involvement of industries in new energy and energy-saving technology (see Figure 4.2). This means that the Torch plan has reached its objective because the plan's initial goal was to get industrial enterprises interested in investing their own funds in the new technology.

Figure 4.1 Funding of National Main Research Program from national budget for energy technology research (excluding government funding for university research).

	2000	**2004**
Total funds arranged (million RMB)	596	2,059
Basic research program[iii]	113	101
863 program	121	1,173
Key technologies R&D program	362	785

Source: Calculated from data in the *China Statistical Yearbook on Science and Technology 2001*, p. 182 and *2005*, p. 359.

[iii] Includes funding for National Natural Science Fund, National Basic Research Program of China (973 program), and National Key Program of Basic Research.

As a result of the ongoing institutional reforms, many government-affiliated research institutes have become independent or been merged with enterprises. These institutes are key players in energy technology R&D. Two such institutes are the China Institute of Power Science that formerly belonged to the Ministry of Energy and is now an enterprise under the State Grid, and the China Coal Research Institute, established in 1957 and turned into a key enterprise in 1999 under the direct leadership of the Central Enterprise Working Commission. Figure 4.3 shows the budgets for energy technologies in these research

Figure 4.2 Funding of national Torch plan dedicated to new energy and energy-saving technology.

	2000	2004
Number of projects	219	308
Fund arranged (million RMB)	2,010	6,709
Government funds		82
Enterprise funds		5,359

Note: In addition to government and enterprise funds, projects receive other funding, for example, projects based on international collaboration.

Source: *China Statistical Yearbook on Science and Technology 2001*, p. 187 and *2005*, p. 360.

Figure 4.3 Research funding in R&D institutions by sectors, 2004 (million RMB).

	Coal mining and processing	Oil and natural gas exploitation	Power generation and transmission	Gasification and supply	Refinery, coking, and nuclear fuel processing
Total	17.9	68.7	326.4	10.8	1.2
Government funding	10.4	5.5	199.1	4.0	1.1
Enterprise funding	1.38	57.6	3.5	1.6	0.1
Others	6.1	5.6	123.9	5.24	0

Source: *China Statistical Yearbook on Science and Technology 2005*, pp. 54–55.

institutes; their funding comes from both the government and industries. The power generation and transmission sectors are the major ones that receive funding because, in the past, the former Ministry of Electricity administered several large research institutes focusing on these sectors.

Research projects typically have mixed funding, with one part coming from the government, another from the enterprise(s) and, in some cases, a third part in the form of bank loans. Industry is increasing its involvement in energy technology R&D, as a result of economic reforms and an increase in its own research demands. Figure 4.4 presents research funding from industries in 2004. The total funding from industries is the major source for energy technology R&D and is expected to keep increasing in the future. This trend is similar to that in developed countries.

4.3 Research institutions and their activities

In China, energy technology research is mainly done in research institutions and enterprises. Historically, the People's Republic of China is a centralized planning nation, and this still applies to current research activities. The government-supported research system includes research institutes in the Chinese Academy of Sciences (CAS) and several ministries as well as research institutes belonging to local (provincial and city) governments. Figure 4.5 presents the framework of energy-related research institutes functioning with government support.

The Guangzhou Institute of Energy Conversion (GIEC), under the Chinese Academy of Sciences, was founded in 1978. In 1998, the institute was chosen as the main institute of high technology R&D of CAS. It has a staff of 236, including 66 senior researchers. The institute is authorized to grant masters and doctoral science degrees. As a state-run institute in the field of clean energy engineering science, the institute's primary task is to conduct research on new and renewable energy utilization technology. It is currently taking the lead in biomass utilization technology, ocean energy development technology, and deep-sea hydrate utilization technology. The GIEC has made advances in such key technologies as the 3 MW Biomass Gasification Power Generation System, the 100 kW sea-wave power generation system, as well as hydrate transport and utilization technology.[10]

Figure 4.4 Research funding in industries by sectors, 2004 (million RMB).

	Coal mining and processing	Oil and natural gas exploitation	Power generation	Power supply	Refinery, coking, and nuclear fuel processing	Power transmission and distribution	Heat production and supply	Gasification and supply	Boiler and original power production	Motor and generator production
Total	5,766.2	4,992.2	2,318.8	2,656.9	4,402.4	2,656.9	179.1	99.6	2,581.4	1,951.3
Government funding	172.7	20.4	41.0	82.1	683.5	82.1	0.4	1.2	63.4	78.3
Enterprise funding	5,350.1	4,197.2	2,222.0	2,532.6	3,414.5	2,532.6	72.2	96.8	2,336.2	1,711.6
Loan from bank	225.1	175.5	31.6	25.9	197.3	25.9	106.5	0.4	172.9	142.9
Other source	18.4	599.0	24.2	16.4	106.1	16.4	0.0	1.1	8.9	18.6

Source: *China Economic Census Yearbook 2004*, p. 8.

Figure 4.5 Leading research institutes in China focusing on energy technology.

Classification	Brief of institute	Leading energy research institutes
Chinese Academy of Sciences	84 research institutes and 24 research centers	Institute of Electrical Engineering Institute of Engineering Thermophysics Institute of Coal Chemistry Dalian Institute of Chemical Physics Guangzhou Institute of Energy Conversion
Institutes affiliated to ministries	Some research institutes belong to Ministry of Transport, Ministry of Agriculture, Ministry of Railway, Ministry of Water Resources	Biogas Science Research Institute Nanjing Agriculture Machinery Research Institute Railway Technology Research Institute Transport Technology Research Institute
Provincial government institutes	Similar pattern to national research organizations, but more focused on local energy technology and renewable energy	Liaoning Energy Research Institute Shandong Energy Research Institute Henan Energy Research Institute
Others	Used to belong to ministries; now belong to enterprises	China Electric Power Research Institute (CEPRI) China Coal Research Institute (CCRI)

The China Electric Power Research Institute (CEPRI), founded in 1951, is both a research institution and a scientific and technological enterprise wholly owned by the State Grid Corporation of China (SGCC). The CEPRI mainly concentrates on pursuing research fields that cover the planning, control, and protection of the power system; high voltage technology; telecommunication and information; transmission and transformation equipment; and new and renewable energy sources. A large number of experts have been trained by CEPRI in the electric power field. The institute boasts one academician from the Chinese Academy of Sciences, one academician from the Chinese Academy of Engineering, and 45 employees with postdoctoral and doctoral degrees.[11]

The China Coal Research Institute (CCRI), established in 1957, is an R&D institute focusing on the Chinese coal industry. In 1999, CCRI

Figure 4.6 Number of researchers in industries in 2004.

	Coal mining and processing	Oil and natural gas exploitation	Power generation	Gasification and supply	Refinery, coking, and nuclear fuel processing	Power transmission and distribution	Heat production and supply	Power supply	Boiler and original power production	Motor and generator production
S&T personnel	91,650	54,172	24,804	2,068	23,367	46,669	1,029	26,997	25,675	22,232
Researchers involved in research projects	55,061	29,321	16,331	1,246	13,852	31,059	669	19,421	15,020	13,825
Administration and service	36,589	24,851	8,473	822	9,515	15,610	360	7,576	10,655	8,407

Source: *China Economic Census Yearbook 2004*, p. 8.

was turned into a key enterprise under the direct leadership of the Central Enterprise Working Commission. There are 17 branches, institutes, centers, and companies in CCRI and they are located in 11 cities. The institute mainly pursues fundamental technology R&D and new technology diffusion on coal mine geography and exploration, coal mine construction, exploitation, safety, machinery, dressing and washing, environment protection, coal liquefaction, and so on. It has made advances in integrated coal mining and exploitation technologies as well as in coal liquefaction technologies.[12]

Shanghai Nuclear Engineering Research and Design Institute was founded in 1970, focusing on research into nuclear power generation system development. The institute conducted the research and design work for the Qinshan Nuclear Plant in China and the Chashma Nuclear Plant in Pakistan. The institute employs more than 600 researchers in eight research branches and five research centers. It belongs to the Shanghai Power Group, which is organized and advised by the Shanghai municipal government that, in turn, is pushing the group to take the lead in nuclear technology development.

Another important nuclear energy research institute is the China Institute of Atomic Energy (CIAE). The CIAE was established in 1950, and conducts research into nuclear physics, reactor engineering, and nuclear technology utilization. Historically, CIAE has been a leading institute on nuclear technology research in China.[13]

4.3.1 Researchers and education

There are nearly 280,000 people working on technology innovation for the energy sectors in industries and national research institutes, according to the National Bureau of Statistics data (see Figures 4.6 and 4.7). There are basically three types of research entities: First, traditional

Figure 4.7 Number of researchers in research institutes in 2004.

	Coal mining and processing	Oil and natural gas exploitation	Power generation	Gasification and supply	Refinery, coking, and nuclear fuel processing
S&T personnel	302	264	1,231	86	10
Scientists and engineers	239	247	1,130	47	8

Source: *China Statistical Yearbook on Science and Technology 2005*, pp. 34–36.

research institutes under, for example, CAS or a university; second, research institutes that function as enterprises; and third, research departments, in some cases very large ones, within an enterprise. From the figures we can see that the number of researchers in industries is considerably larger than that in research institutes.

Though the number of researchers within enterprises working on energy-related research is substantially larger than the number in research institutes, it is worth remembering that many researchers within enterprises mainly work on product development. Moreover, the figures that provide data on researchers working within industry do not include researchers working in energy areas that belong to the machinery-manufacturing sector. The dominance of coal is also evident when assessing researcher data, although figures are still small when compared to the more than five million people employed overall in the coal industry nationwide.[14]

Important energy technology R&D is also being carried out in universities. A case in point is the advanced thermal power system research being conducted at Shanghai Jiaotong University and Zhejiang University. Additionally, the Thermoenergy Engineering Research Institute of Southeast University joined the magnetohydrodynamic (MHD) power generation research project under the 863 program, while Jilin University and Xian Jiaotong University joined the alternative fuel on ethanol combustion key research project within the 973 program.

Figure 4.8 shows the energy technology R&D programs, expenditure, and staff in universities. Energy technology R&D in universities mainly focuses on basic research. Cooperation with enterprises has become more common of late.

China has numerous universities that concentrate solely on energy-related education, such as the China Petroleum University, the North China Electric Power University, China University of Mining and Technology, and so on. Funding for these universities comes both from the government budget and from industries. Most of the key universities have energy-specific departments or institutes. For example, Tsinghua University's Department of Thermal Engineering includes the Energy Power System, Institute of Nuclear and New Energy Technology; Shanghai Jiaotong University has the Energy Research Institute, School of Nuclear Science and Engineering; and Zhejiang University has the College of Mechanical and Energy Engineering.

Figure 4.8 Energy-related R&D programs, expenditure, and staff in institutes of higher education.

	Number of projects		Scientists and engineers		Expenditure (million RMB)	
	2001	2004	2001	2004	2001	2004
Power and electrical engineering	3,334	4,577	3,168	3,927	444	639.9
Energy technology	810	1,237	860	938	66	186.5
Nuclear technology	158	207	156	204	115	103.6

Source: *China Statistical Yearbook on Science and Technology 2002*, p. 378 and *2005*, pp. 348–349.

4.3.2 International collaboration

There are many kinds of international collaboration aimed at enhancing energy technology innovation in China, both within research institutes and industries. In general, there are three types of collaboration between China and other countries: government-supported activities, collaboration between research institutes, and industry collaboration.

Government-supported collaboration is mainly either multilateral or bilateral collaboration on technology development. The MOST and other relevant ministries coordinate numerous national collaboration programs, several of which include energy technology R&D. China–EU Science and Technology Cooperation and China–EU Partnership on Climate Change are the main programs between China and Europe. The former is an overall S&T program that led to a memorandum of understanding in 2005 between China and EU on transport and energy strategies, including, for example, clean coal technology.[15] Clean coal technology is also an important factor in the latter program that focuses on low-carbon technologies, and which emphasizes renewable energy technologies and energy efficiency, as well as a range of near zero-emission coal technologies such as carbon capture and storage.[16] However, concrete projects are yet to materialize.

Clean coal technology development and transfer are even more important in the Asia–Pacific Partnership for Clean Development and

Climate (APPCDC),[iv] which is reluctant to impose Kyoto Protocol-type restrictions on greenhouse gas emissions. The APPCDC focuses on cleaner fossil fuel, renewable energy and distributed energy supply, coal mining, power generation and transmission, as well as energy efficiency.

In order to promote international technology R&D collaboration, MOST has set up the International Science and Technology Collaboration Plan. There are several programs in the plan, including the International Key Science and Technology Collaboration Plan, the EU Collaboration Plan, the China–Australia Young Researcher Exchange Plan, and so on. In the Eleventh Five-Year Science and Technology Development Plan, energy technology and water resource protection technology are the first two sectors targeted for international cooperation, highlighting their importance.[17] However, it is too early to assess whether the results from these numerous collaboration projects will lead to groundbreaking research.

Collaboration between research institutes in China and other countries is another important way to promote technology R&D collaboration. International collaboration is the main vehicle for China to develop emerging technology R&D.

One well-known form of international collaboration is the project to build the International Thermonuclear Experimental Reactor (ITER), said to be the world's most advanced nuclear fusion reactor. China will pay 10 percent of the total budget for this project. Chinese participation consists mostly of specialists from the Institute of Plasma Physics of CAS and South Western Institute of Physics, Nuclear Fusion Theory and Simulation Center of Zhejiang University. In 2006 the China Institute of Atomic Energy announced that it had made significant progress in controlling the high-temperature reaction process, which could be a key element in extending the control time for the fusion process.[18] Controlling the process, as well as the immense heat needed during it, are the main stumbling blocks the scientists have to overcome before ITER is ready.[19]

Another example of international collaboration is the joint work carried out by the Guangzhou Institute of Energy Conversion and Nagoya University, as well as Toyota Automobile Co. on utilization of

[iv] Also called AP6 since there are six member states: Australia, China, India, Japan, South Korea, and the United States.

landfill. Waste-derived fuel is an important way to efficiently utilize landfill. This study focuses on adding stuffing during the moulding process, to reduce pollution in furnaces and reduce end-pipe emissions. This project will develop new theory and technology for waste utilization and establish an advanced research base in China.

Shandong University is working together with the Swedish Royal Institute of Technology on a heat-exchange platform, by studying heat-exchange simulation and by identifying parameters, design methodology, and so on. This project is supported by the Key International Collaboration Project in MOST.[20]

However, in the field of energy technology, industry collaboration is still the major way for China to localize advanced technology. Most of the important energy technology manufactured in China is learned from other countries. China often uses the strategy called "Use the market to get technology," which is also supported by the government. For example, China has acquired windpower technology, large hydropower technology and natural gas combined-cycle technology through industrial collaboration. Recently there has also been significant progress in technology collaboration through commercial transactions. Chinese industrial companies have bought foreign technology that is manufactured in China.

During the Three Gorges Dam project, companies made bids for the hydropower generation contract, and technology transfer was one of the prerequisites of the bidding. Dongfang Electrical Machine Company (DFEM) and Harbin Power Equipment Company (HPEC)[v] were partners in China for technology transfer from Alstom and VGS. As a result of the technology transfer during the project, the two Chinese companies can now manufacture 700 MW hydropower generators.

In 2000, China started a natural gas power-generation program and invited foreign manufacturers to participate in the bidding. As with the Three Gorges Dam project, technology transfer was a requirement. In 2004, HPEC worked together with General Electric (GE) and DFEM with Mitsubishi and Nanjing Turbine to produce large natural gas turbines in China. By 2005, the Chinese companies could already produce large natural gas turbines with more than 60 percent of the components manufactured domestically.

[v] Harbin Power Equipment Company Ltd is also variously referred to as Harbin Power, Harbin Power Plant Corporation, Harbin Power Engineering Company Ltd, and Harbin Power Equipment Group Corporation.

4.3.3 Energy-related publications

The number of Chinese energy-related publications has risen dramatically in recent years. Each area of energy technology development has some outstanding journals, specified in Figure 4.9.

Figure 4.9 Major journals related to energy technology in China.

Sectors	Leading journals	Notes
Power generation technologies	中国电力 China Power 电工技术 Electric Engineering 电网技术 Grid Technology	Focus on power generation technology, transmission and distribution technology innovation. Publications on key and advanced technologies in power projects, national programs, together with national policies. These journals are key national journals and on the list of Engineering Information (EI).
Coal mining technology	煤炭科学技术 Coal Science and Technology	Important journals in coal mining industry, with articles on technology development and innovations.
	煤炭技术 Coal Technology	Both are key national journals and on the list of EI.
Oil and natural gas exploitation	石油工程建设 Petroleum Engineering Construction 石油钻探技术 Petroleum Exploration Technology	Publication on oil reserves, exploration, exploitation technology, important projects, and key research. It is a key national journal and on the list of EI.
Renewable energy	太阳能 Solar Energy 风力发电 Wind Power Generation 新能源 New Energy	All three are important journals on renewable energy development, including technology development.

4.3.4 Companies and their research activities

Thanks to the reforms initiated by the Chinese government, the energy technology research capabilities of state-owned companies are stronger today than they had been in previous decades. There are many research institutes that used to belong to the former Ministry of Electricity, Ministry of Oil or Ministry of Coal, which are now administered by state-owned enterprises. Private companies have also begun to play a role in technology development, in a similar way as in other countries.

The rapid growth in the demand for energy has provided many opportunities for energy technology manufacturers in China (see Figure 4.10). For the past few years, newly installed power-generation capacity has been around 40 to more than 100 GW/year, and is expected to be at least 70 GW in 2007. The HPEC, DFEM and Shanghai Electric are the top three Chinese suppliers for coal-fired power generation equipment. Coal production has increased from 1.28 billion tonnes in 2000 to 2.14 billion tonnes in 2005.[21] This increase in coal use, together with new coal-mining safety requirements, has spurred demand for new coal-mining equipment. According to the Eleventh Five-Year Plan for the coal industry, 95 percent of the large and 80 percent of the medium-sized coalmines[vi] must be equipped with modern machinery by 2010.[22] Owing to the high price of oil, investment in domestic oil resources has also increased rapidly. In 2005, investment in oil and natural gas exploitation was nearly RMB 150 billion (USD 19 billion), 30 percent higher than in 2004.[23] This gives Chinese energy technology manufacturers incentives to extend production capacity.

At the same time, Chinese energy technology manufacturers have to try to keep their technology as advanced as possible, by learning from companies in other countries and by fostering their own technological innovations. Learning from other countries is the key though some large manufacturers have recognized the importance of the ability to create their own technology R&D and have started to establish in-house research teams. However, compared with companies in other countries, enterprise investment in technology R&D is still quite limited in China. Of the 103 Chinese companies that qualify as technological innovators, according to guidelines drawn up by MOST on the basis of the percentage of sales invested in R&D, only nine are companies dealing with energy-related technology.[24]

4.4 Energy-related R&D activities in China

The main factors driving energy technology R&D in China are related to supply, demand, and competition. An abundance of domestic

[vi] There are about 28,000 coalmines in China. About 24,000 mines are small and most of them are not mechanized. The government has drawn up plans to substantially reduce the number of small mines by means of mergers. See "专访王德明教授：煤矿安全呼唤科技保障" [Interview of Professor Wang Deming: S&T safeguarding needed for the safety of coal mines], *Renmin Ribao* [People's Daily], 26 October 2004.

Figure 4.10 Important companies in major areas of energy technology development.

Area	Companies	Brief
Power generation	Harbin Power Equipment Company Ltd (HPEC) Dongfang Electrical Machine Company Ltd (DFEM) Shanghai Electric Group Company Ltd	Key power generation suppliers in China for coal-fired power plants, natural gas power plants, hydropower, and nuclear power technologies. With government support for international technology collaboration, these companies have become some of the leading technology manufacturers in the world.
Wind turbine	Xinjiang Jinfeng Co. Longyuan Wind Co. Harbin Power Zhonghang Huiteng Co.	Leading manufacturers in China for wind turbines. Xinjiang Jinfeng is a private company, others are state-owned.
Biomass gasification	Hongyan Machinery Co.	Private company, works together with Guangzhou Institute of Energy Conversion in technology development.
Coal gasification	Taiyuan Coal Gasification Group (TCGG) Yanzhou Coal Group (YCG)	The TCGG is the first company in China to implement integrated coal utilization including coal gasification. Formed by the former coal ministry and Shanxi Provincial Government, it has its own research center. The YCG is becoming a leading company in China for underground coal gasification. It has its own research center and own gasification technologies.
Coal liquefaction	Shenhua Group (SG) Shanxi Coal Group	The first direct coal liquefaction production is being implemented by the SG in Inner Mongolia in China. Based on collaboration with CAS, SASOL, and Shell, another two coal liquefaction production plants in Ningxia and Shanxi will undergo construction in 2007.
Coal mining	Taiyuan Heavy Machinery Group Co., LTD	Largest manufacturer of coal-mining systems in China. Has its own technology research center with a staff of 410. Is one of 103 enterprises chosen for technology innovation pilot projects.

Continued

Figure 4.10 Continued

Area	Companies	Brief
Oil exploitation	PetroChina	Is becoming one of the world's leading companies for exploration and exploitation technology. In 2006, it was awarded the World Oil Prize for new technology.
Solar PV	Shangde (Suntech)	Leading solar PV technology manufacturer and research company. Private holding.

energy resources, especially coal, and the large demand for energy in China have created a market for energy technology manufacturers, producing items ranging from power equipment and boilers to mining equipment and energy conversion technology. It has also spurred Chinese companies to become internationally competitive, which is necessary not only for them to succeed in the domestic market, but also abroad. Major power companies now have their eye on the international market. The Chinese government considers technology innovation to be one of the major measures to improve competition. Since the beginning of the Eleventh Five-Year Plan, a company's five-year plan has had to include investment in technology innovation, especially in state-owned large manufacturers and sector associations.

Differences in the energy utilization system in China and many other countries could create advantages for Chinese manufacturers. For example, if coal use in other countries decreases, their technology demand would also decrease, and as a result foreign manufacturers would face difficult times. This in turn would probably lead to a decrease in investments by foreign companies in technology R&D, as recent trends have shown. Meanwhile, Chinese investment in coal utilization technology, hydropower, nuclear power, and biomass utilization technology is on the rise. As a result, Chinese technology can be expected to advance to a globally competitive level in the future.

4.4.1 Clean coal

Clean coal technologies will have the most promising future in China, as the country is becoming the world's largest market for coal-fired power generators. Basically, clean coal technology (CCT) covers technologies that reduce pollution and increase efficiency, including coal processing, combustion, conversion and pollutant control.[vii]

According to the Chinese definition in "China's Ninth Five-Year Plan and 2010 Development Outline for Clean Coal Technology," CCT includes 14 technologies in four areas, as shown in Figure 4.11.

Three major Chinese power equipment manufacturers (HPEC, DFEM and Shanghai Electric) have already mastered supercritical unit technology and are advancing toward technology needed in ultra-supercritical (USC) power units. There are more than ten USC units under construction, most of which are domestically manufactured. Another ten or so units are scheduled for construction in 2007. The dominance of supercritical and USC units will inevitably weaken the

Figure 4.11 Clean coal technologies listed in China's Ninth Five-Year Plan for clean coal technology.[25]

Coal processing	**High-efficient coal combustion and advanced power generation**	**Coal conversion**	**Pollutant control and waste processing**
Coal dressing	Ultra-supercritical (USC)*	Gasification	Fluid gas emission control
Industrial briquette	Circulating fluidized bed combustor (CFBC)**	Liquefaction	Utilization of powder coal ash
Coal water mixture	Pressurized fluidized bed combustion (PFBC)**	Fuel cell	Utilization of coal bed methane
	Integrated gasification combined cycle (IGCC)		Utilization of coal stone
			Coal washing water

Notes: * USC is moving into the commercialization stage in China and was not included in the original plan.

** The technology has already progressed to the commercialization stage though dissemination of both CFBC and PFBC has been limited.

[vii] One step in CCT has been to improve the efficiency and reduce the emissions of power plants by increasing the boiler temperature above the critical point where liquid turns into steam. This is known as the supercritical level. The ultra-supercritical level means that the boiler temperature is even higher and therefore clearly above the sub-critical point.

position of other coal-fired power technology. Chinese manufacturers are being propelled toward new technologies for USC power generators by sheer demand and the market, and in the process they are becoming the most advanced in the world. The Chinese government also supports large manufacturers to extend their business and enhance their technology innovation.

According to the document "Guidance for Restructuring the Energy Industry," issued by the National Development and Reform Commission (NDRC) in 2005, thermal power generation units of less than 100MW are to be limited and should be removed and replaced by larger units, except in cases of cogeneration[viii] for the purpose of producing electricity and heat for industry and buildings.[26]

More than 80 percent of China's power energy capacity increase in recent years has come from coal-fired power plants,[27] and more than 80 percent of newly installed coal-fired power generators are large units with 300 MW or higher capacity. In 2005, there were around 110 coal-fired power generators of 300 MW or higher with newly installed capacity of nearly 45 GW. China produces about half of the world's newly installed coal-fired power capacity.[28] The efficiency of a USC unit could be 43 percent in the near term, at a relatively low cost of around RMB 5000 /kW. In the long term, the efficiency could reach 46 percent.

For a long time, there was controversy regarding the technology development pathway for coal-fired power generators; a decision could not be reached on whether to advance toward USC units or the Integrated Gasification Combined Cycle (IGCC). It seems that Chinese manufacturers used to have a preference for USC units simply because they are close to having the capacity to produce them, in addition to their being cost-effective. The USC unit technologies are being manufactured in China using technology originating in foreign countries, with 80–90 percent of the components being produced domestically. The HPEC, DFEM and Shanghai Electric are already the three biggest manufacturers of coal-fired power units in the world. The Yuhuan Power Plant, constructed in 2006 by the Huaneng Group, was the first to use domestic technology, provided by Harbin Power

[viii] Cogeneration means simultaneous creation of heat in addition to the electricity produced in a power plant. The heat can be used for various purposes, making the overall energy use very effective.

and Shanghai Power. In recent years 300 MW to 600 MW units have constituted the major coal-fired power units installed in China.

Despite, or perhaps even because of, the strides that have been made by Chinese companies in mastering USC unit technologies, it is now evident that major power companies have shifted their focus to IGCC and want to become the world leaders in this cutting-edge technology. The IGCC is a new technology that is still mainly at the developmental stage. The efficiency of modern power stations can exceed 40 percent (Lower Heating Value) although the average efficiency worldwide is about 33 percent. The IGCC systems utilize efficiency and low capital cost advantages of a Combined Cycle Gas Turbine (CCGT) by first gasifying coal or other fuels. Gasifiers are usually oxygen blown and are still at an early commercial stage. Coal and difficult liquid fuels such as bitumens and tar can be used as feedstock. The potential efficiency of IGCCs is around 51 percent, based on the latest CCGTs of 60 percent efficiency.[29] With continuing development in hot gas cleaning and better heat recovery as well as the continuing development of CCGTs, commercially viable coal- or wood-fired IGCC power stations with efficiencies of over 60 percent may be feasible by 2020.

In addition to the potentially high efficiency, IGCC offers one of the more promising routes to CO_2 capture and disposal by converting the gas from the gasifier into a stream of H_2 and CO_2 via a shift reaction. The CO_2 can then be removed for disposal before entering the gas turbine. The resulting stream of H_2 could be used in fuel cells and not just in a gas turbine. The IGCC technology can also be used for small-scale plants without sacrificing efficiency. Project feasibility studies were conducted for the Beijing IGCC project and Yantai IGCC project between 1995 and 2001, with partial funding from the World Bank.

The Yankuang Group and East China Polytechnic University jointly undertook a project for IGCC, together with a methanol generation system. Methanol, generated through the coal-to-liquid method, can be added to gasoline to make cleaner fuel for vehicles. This project got underway in 2003 with funding from the 863 program. The ethylic-acid-based cogeneration project was first tested in July 2005, and in May 2006 it was approved for further use.[30]

The IGCC is on the government's list of key technologies to be promoted, with strong government support. The IGCC technology is an important part of the national technology R&D programs such as the

863 program. Chinese manufacturers can basically produce most parts of IGCC system, but they still lag behind in mastering the integration process and compilation of the more advanced parts.

With the price of energy rising and China's national strategy focusing on sustainable development, polygeneration is receiving a lot more attention (see separate side-bar). A preliminary analysis of polygeneration was conducted within the 973 program, and several companies, research institutes, and universities are planning to construct polygeneration systems, including coal gasification power generation, fuel, and feedstock production. The Yanzhou Mining Co. is working with the Engineering Thermal Physics Research Institute of CAS on a pilot project. This project will have 76 MW power generation and 240 kilotonnes methanol production. The Shenhua Group is on the lookout for a project on coal liquefaction and power generation. The Shanxi Coal Chemistry Institute of CAS, Eastern China University of Technology, National Coal Water Mixture Gasification, and Coal Chemistry Engineering Center are leading research institutes when it comes to this technology. A number of polygeneration projects with output on power generation, methanol, and dimethyl ether (DME, another clear-burning fuel) are under design and construction in China, and will be completed in the time frame 2007–09.

POLYGENERATION—A COMPREHENSIVE SOLUTION

Zheng Li

Polygeneration can greatly increase coal use efficiency and consequently alleviate the pressure on China to meet the demand for resources. Overall energy efficiency can be increased through energy coupling in the cogeneration process of power and liquid fuels on a large scale. At the same time, because of the economies of scale brought about by the public utilization of a gasification device in power generation and chemical production, costs can be reduced. China has an abundance of coal resources with a high sulphur content, and if the coal can be used as the raw material for polygeneration at a relatively low cost, the product cost can be reduced further. Therefore, the application of an efficient and economic polygeneration system in coal-rich areas, particularly those with a high sulphur content, will be very helpful in meeting China's future energy requirements.

To ensure oil security in China, it will be necessary to find ways of developing alternative liquid fuel, with coal being the first choice in this respect in large-scale production. Polygeneration can cogenerate power and a range of fuels, including methanol, dimethyl ether, F-T liquids, and pure hydrogen. Low-cost liquid fuels produced by Polygeneration systems can compete with petroleum and diesel as a viable alternative. It will not only alleviate the pressure to import oil, but will also help to ensure China's energy security in the long term.

Based on a modular structure, the polygeneration system is integrated with different traditional technologies and production processes. Figure 4.12 shows the conceptual schema of the system. The basic process is fuel gasification through a large-scale gasifier, in which the fuel includes coal (with a particularly high sulphur content), petrol coke, biomass, and so forth. After gas purification processes, the clean syngas can be converted by a once-through process to clean fuels such as methanol, dimethyl ether, F-T liquids, or other high value-added chemical products. The un-reacted tail gas can be directly transferred to the power generation sector to produce electricity. If necessary, the syngas can be converted to hydrogen and carbon dioxide through a water shift reaction; then after separation, hydrogen can be used for centralized electricity production or decentralized fuel cell vehicles in the future. The concentrated carbon dioxide captured from the system can be utilized in the form of chemical raw materials, EOR, ECBM, and so on, or in direct sequestration underground. Based on this idea, the polygeneration system can be configured using various

Figure 4.12 Conceptual diagram of polygeneration.

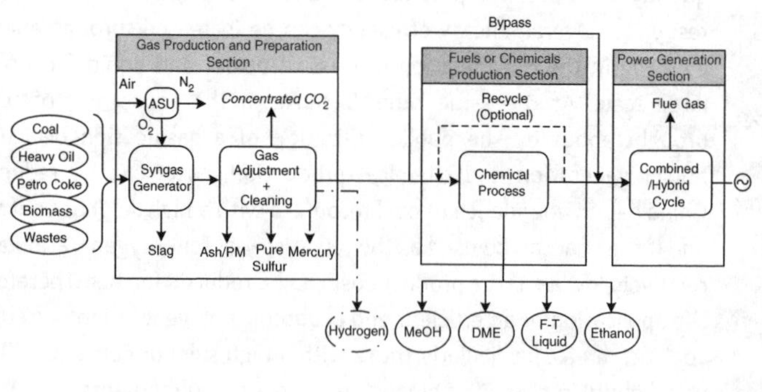

coupling methods. As a result, it can meet different requirements based on variations in the geographical distribution of resources, varying strategic needs over the course of time, and diverse stages of technology development.

Environmental concerns must be addressed both now and in the future. Bearing this in mind, conventional pollutants such as SO_2, NO_X, and PM_{10} created in the production processes of the polygeneration system based on gasification technologies are on par with those of natural gas power plants, which means that polygeneration can resolve the serious pollution in coal-fired power plants. Additionally, SO_2, CO, NO_X, PM, and other pollutants caused by the combustion of liquid fuels (e.g., methanol and dimethyl ether) in the internal combustion engine are significantly lower than current gasoline and diesel vehicles, which will naturally improve the quality of the urban environment.

High CO_2 emissions are an inevitable consequence of coal utilization. Apart from reducing total carbon emissions by improving efficiency to decrease coal use, the capture and sequestration of CO_2 is the only way to reduce CO_2 emissions in the long term. Advanced coal-combustion technology will lead to high efficiency, but it is difficult and expensive to capture CO_2 from the flue gas because of its low concentration. Polygeneration has particular advantages when it comes to CO_2 capture because, after the water gas shift reaction and separation processes, there is a high concentration of CO_2 in the syngas, which is easier to extract. At the same time, pure hydrogen can be utilized in the future hydrogen economy.

To summarize, to achieve the sustainable development of China's energy system, polygeneration is both a comprehensive solution with remarkable prospects as well as a bridge between mid- and long-term development. In the near-term future, polygeneration can be used to produce power and oil-alternative products in a more efficient, economic, and clean way, which will alleviate the national energy shortage. In the long-term future, CO_2 can readily be captured by polygeneration, which will help to meet the anticipated demands for CO_2 reduction in China. Simultaneously, pure hydrogen can be produced with CO_2, which will accelerate the transition to a hydrogen economy in China.

Professor Li of Tsinghua University is also Director of the Tsinghua-BP Clean Energy Research and Education Centre.

4.4.2 Nuclear

Owing to energy supply pressure, environmental problems and increased manufacturing capacity in China, nuclear power development has become a major feature of the national energy development strategy. The 863 program also supports nuclear energy. According to the "Nuclear Power Medium- and Long-term Development Plan (2005–2020)," published by NDRC in 2005, total nuclear power generation should be 40 GW by 2020, while another 18 GW will be under construction at the same time.[31] As a result of this goal, China is currently the world's biggest market for nuclear power equipment manufacturers. Shanghai Electric and DFEM have taken the lead among domestic manufacturers.

Due to the heavy involvement of the world's leading manufacturers, technology transfer to China is a very important component in nuclear technology manufacturing in the country. To be permitted to participate in the construction of nuclear plants in China, companies from other countries are required to transfer technology there. There will be strong collaboration between Chinese and foreign companies on future projects: Shanghai Electric has in the past teamed up with Westinghouse, while DFEM has worked together with Alstom. In December 2006, Westinghouse won the bid to build four new third-generation nuclear plants in China, committing to significant technology transfer.[32]

The Chinese government has made plans for domestically owned technology development. The DFEM, working together with the China Institute of Atomic Energy (CIAE), built the latest Qinshan nuclear power plant, mainly constructed with domestic technology. Qinshan 2 was named one of China's top ten S&T advancements in 2004. The CIAE is also collaborating with Shanghai Electric on the construction of the Lingao Nuclear Plant, with the French Areva NP as a partner. Areva NP is currently constructing what will be the world's largest nuclear reactor, Olkiluoto 3, with a 1.6 GW capacity, in Finland. The problems Areva has had in Finland, which will delay the planned completion of the plant, could have affected the Chinese decision to favor Westinghouse over Areva NP.[33]

All three major Chinese companies have their own research teams working in tandem with Chinese research institutes. Domestic technology already constitutes three quarters of the technology needed for a nuclear power plant. Chinese companies are expected to fully manufacture domestic nuclear power generation equipment in the near future, both with their own technologies and via technology transfer from abroad. Owing to the strong demand for nuclear equipment to support domestic nuclear development, Chinese companies could well become the leading manufacturers in the world during the coming decades.

When examining nuclear power station construction in China, it seems that the country has tried various nuclear technologies through collaboration with different countries. China has yet to publicly make a fixed nuclear technology choice, though Westinghouse's success in securing an order of four plants seems to indicate that the country has decided to favor Westinghouse's AP1000 pressurized water reactor technology. As a result of the rapid expansion of nuclear power generation during the coming decades, China should opt to avoid wasting resources by experimenting with a wide range of technology, and also to simplify the nuclear power generation system.

In order to develop new nuclear technologies, Chinese researchers are working on a pebble-bed high-temperature reactor, an experimental fast reactor, and nuclear fusion technologies under the framework of the 863 program. Tsinghua University's Nuclear and New Energy Research Institute constructed the first 10 MW high-temperature reactors in the world in 2000. An experimental fast reactor with a capacity of 65 MW is also under construction and should start operating in 2010. This marks the first step toward fast reactor development in China. The following two steps include the construction of 300 MW and 1000–1500 MW reactors in the future, based on collaboration between China and Russia. As far as nuclear fusion is concerned, China is collaborating with several other countries in the previously mentioned ITER project.

4.4.3 Renewable energy

Due to the rapid growth in the demand for energy, coupled with increasing environmental problems, renewable energy has received a great deal of attention. According to the Chinese government's plan, it should account for 15 percent of total primary energy use by 2020 (including large hydropower).[34] Considering that renewable energy accounted for about seven percent in 2005, this is an ambitious plan.

Should it materialize, China would become one of the world's leading countries in the use of renewable energy. In 2005, China and Germany were the world's top investors in this respect. China's share of the world total (USD 38 billion) was 7 billion, but the country also had the biggest existing capacity, especially for small hydropower and solar hot water, as well as the biggest increase in renewable energy capacity (7 GW) in the large hydropower plants.[35]

Many Chinese research institutes, of which Guangzhou Institute for Energy Conversion is the leading one, are now working on biomass utilization technology. Advanced biomass utilization technology is also a component of the 863 program. Biomass gasification power-generation technology is ready for commercial use and direct biomass combustion boilers are being produced in China. The Guanghua Company will provide boilers for the Jinzhou Straw Power Plant, which is one of three pilot projects for biomass power generation. In Hainan and Fujian provinces, 1.2 MW units have already been installed. However, compared to advanced technology in other countries, such as BWE in Denmark, the system in use in China still lags behind in efficiency.

In addition to power generation technology, biomass gasification for village use has been rapidly implemented in China. According to the 2006 "New Socialist Countryside" program, an increasing number of villages will install biomass gasification systems for coking, hot water, and space heating, with support from the government.[36]

Though windpower and hydropower generation technology in China has lagged behind the more developed countries, in recent years there has been some development in both fields. One reason for this has been government regulations that have resulted in active technology transfer to China. For example, in 2002, the government issued a regulation stipulating that at least 70 percent of components in installed windpower generators should be manufactured in China. However, this has mainly had an impact on the manufacturing sector instead of advancing R&D, and domestic research on wind turbines in China is not very substantial. For example, one of the key windpower generator manufacturers in China, Xinjiang Jinfeng Co. has only seven researchers.[37] Moreover, the largest windpower generators in China are 1.2 MW units, manufactured by HPEC, while in other countries there are already generators with 3.5 MW capacity. Additionally, offshore wind farms are still in the preliminary stages of development in China.

When it comes to hydropower, the Three Gorges Project has created several opportunities for technology transfer. In line with projects involving nuclear technology, foreign companies aiming to bid for the equipment supply contract had to sign a contract committing the transfer of technology to their Chinese partners. According to the contract, at least 30 percent of the components were to be produced by Chinese manufacturers and two units were provided with major components produced by Chinese manufacturers. The HPEC worked together with Alstom for eight units, while DFEM collaborated with GE and Siemens for another six units. A further 32 units larger than 700 MW are planned for the upstream of the Yangtze River, which will provide good opportunities for Chinese manufacturers. Together with hydropower generators, several companies in China are also developing a hydropower plant control system based on domestic technology.

Chinese research institutes are conducting research on solar energy utilization technology development, both for solar photovoltaics (PV) and solar thermal power generation. The Institute of Optics and Electronics of CAS, Changchun Institute of Optics, Fine Mechanics and Physics, and Shanghai Institute of Optics and Fine Mechanics are the leading bodies in this respect. Working together with these research institutes, Shangde (Suntech Power) installed solar PV production lines for 120 MW in 2006. Shangde has become one of the world's leading solar power companies, with fully owned technology and 98 percent of its products being exported. In fact, Shangde is a prime example of a Chinese high-tech company whose origins can be traced to engineers who returned to China after receiving their higher education abroad. The company's leading figures include Shi Zhengrong, Ji Jiangji, and Stuart Wenham, all of whom collaborated previously at the University of New South Wales and at a company producing solar panels in Australia. Shangde invested in its R&D capabilities in 2006 by purchasing MSK Corporation, a Japanese solar cell company. Suntech's net revenues have soared and totaled USD 148.5 million in the third quarter of 2006. In addition to working with domestic research institutes, Shangde has a collaborative research agreement with the University of New South Wales.[38]

The CAS has also conducted research on solar thermal power generation technology and has made significant progress with regard to a power control system that may result in reduced costs. A smaller solar thermal power generation pilot plant was constructed in Jiangsu province, producing electricity at a cost of RMB 1/kWh. A further three large-scale solar thermal power generation plants will be constructed before 2010.

4.4.4 Others

China's natural resources and energy security have also created a need to develop more advanced methods in the exploration of fossil fuels such as coal, natural gas, and oil. At the moment these technologies are mainly imported to China, although some indigenous R&D is slowly taking place, mainly in large companies such as Xian Coal Machinery, Sinopec, and PetroChina. PetroChina's R&D expenditure (USD 445 million in 2005) is the largest of all Chinese industrial companies, although it is universally difficult to make a distinction between expenditure that goes to the development or maintenance of exploration technologies and hard core research expenditure.[39]

The development of fuel cell and hydrogen technology is also supported by the government. Substantial investment has recently been made in fuel cell R&D. The internal fuel of fuel cells is hydrogen that can also be generated from several other sources, for example methanol and gasoline. Major research units associated with Dalian Institute of Chemical Physics of CAS, Shanghai Tongji University, and Shanghai Fuel Cell Vehicle concentrate on fuel cell. The fuel cell car prototype "Chaoyue 2," produced by Tongji University, represents one of the most advanced technologies in the world. Although it is technologically still far from being a real car ready for mass production, it is the result of truly independent research work being conducted in China, and is progressing at almost the same pace as in other countries.

4.5 Conclusion: Level of energy research in China

Alongside the rise of GDP in China, investment in science and technology, including investment in energy technology R&D, has increased significantly. Technology innovation has been recognized as a key factor for future social and economic development by the Chinese government and Chinese enterprises.

There are many research institutes in China focusing on energy technology R&D. These institutes mainly work on fundamental energy technology R&D and engage in considerable international collaboration. Research institutes are working together with industrial enterprises in order to facilitate energy technology transfer to the manufacturers at the local level. This technology transfer will be instrumental for the next generation of energy technology.

As a result of institutional reform in China, energy technology R&D activities are transferring to enterprises. As more and more enterprise-owned research teams are established, the manufacturing of advanced energy technology in China is improving. Major manufacturers of energy technologies have also been able to receive transferred technology with governmental support.

Although Chinese energy technology manufacturers are getting stronger, their innovation ability is still very weak and their own investment in technology R&D quite small. Even though Chinese companies have absorbed technology know-how from foreign companies, there are still concerns about keeping abreast of global technology innovation. There is always a danger that Chinese manufacturers will continue the trend of "lag behind—learn technology—lag behind again—learn technology again," if they do not have the ability to innovate independently. For this reason, this is a good time for further investments in technology innovation.

China's very high saving rate (46 percent in 2005)[40] provides many investment opportunities. Large amounts of private investment have gone into and still go into energy-intensive industries, such as the production of cement and steel, which is damaging for the overall economic structure. It would be better to increase investments in technology innovation to facilitate the creation of a society based on sustainable development.

In the area of energy technology, China does challenge the world in some fields. Owing to the country's large energy demand and coal supply, it may, during the next decade or so, take the lead in clean coal utilization technology such as advanced coal-fired power generation, coal gasification technologies and coal liquefacation technologies. China is also making a concerted effort to improve technology development in nuclear power generation and large hydropower generation technologies. This not only contributes to the country's economic growth and energy security, but is also beneficial for the global environment and global technology development.

Even though there has been a great deal of progress in the field of energy technology development in China, many energy technologies in the country are still lagging behind developed countries; for example, IGCC and near-zero-emission power generation technology, advanced natural gas combine cycle (ANGCC), solar PV, and solar thermal power generation. In addition, China's technological level is not up to the standard of developed countries in such areas as advanced biomass liquefaction technologies, offshore windpower generation, oil and natural gas reserve exploitation technologies, and fuel cells for power generation. These technologies are more future-oriented ones, and need further research support from the government. Moreover, China is weak in emerging technologies such as unconventional energy utilization and advanced nuclear technologies.

In conclusion, China's growing energy demand will continue to put enormous pressure on the environment and on the country's energy supply. If China manages to find solutions to make new, clean and high-efficiency energy conversion technologies commercially viable, China will not be the sole beneficiary. The energy outlook for the world would look brighter too.

Notes

1. K.J. Jiang and X.L. Hu, "Energy Demand and Emissions in 2030 in China: Scenarios and Policy Options," *Environment Economics and Policy Studies*, 8 (2006), pp. 129–141.
2. National Bureau of Statistics of China, *China Statistical Yearbook 2006* (Beijing: China Statistics Press), p. 261.
3. International Energy Agency IEA, "The World Energy Outlook 2006 maps out a cleaner, cleverer and more competitive energy future," 7 November 2006, http://www.iea.org/Textbase/press/pressdetail.asp?PRESS_REL_ID=187.
4. K. Bradsher and D. Barboza, "Pollution from Chinese coal casts a global shadow," *The New York Times*, 11 June 2006.

5. F.C. Li, "Green goal missed by big margin," China Daily, 10 January 2007, http://www.chinadaily.com.cn/china/2007-01/10/content_779106.htm.
6. State Council of the PRC, "国家中长期科学和技术发展规划纲要" [Outline of National Medium- and Long-term S&T Development Plan (2006–2020)], 9 February 2006, http://www.gov.cn/jrzg/2006-02/09/ content_183787.htm.
7. State Council of the PRC "加快振兴装备制造业的若干意见" [Several suggestions for accelerating equipment manufacture], 13 February 2006, http://www.cas.cn/html/Dir/2006/06/29/14/16/66.htm.
8. For more information on the 863 program see, http://www.863.org.cn/english/index.html.
9. For more information on the 973 program see, http://www.973.gov.cn/English/Index.aspx.
10. For more information see the GIEC website, http://www.giec.ac.cn/giec/giec_english/about%20giec/about_introduction.htm.
11. For more information see the website of CEPRI, http://www.epri.ac.cn/en/index_E.jsp.
12. For more information see the website of CCRI, http://www.ccri.com.cn/e_index.asp.
13. For more information see the website of CIAE, http://www.ciae.ac.cn/02gk/jj.asp.
14. National Bureau of Statistics of China, *China Statistical Yearbook 2006* (Beijing: China Statistics Press), pp. 506–507.
15. European Commission, "An overview of sectoral dialogues between China and the European Commission," http://ec.europa.eu/comm/external_relations/china/intro/sect.htm.
16. European Commission, "EU and China partnership on climate change," 2 September 2005, http://europa.eu.int/rapid/pressReleasesAction.do?reference=MEMO/05/298.
17. Ministry of Science and Technology, "国家'十一五'科学技术发展规划" [Eleventh National Five-Year S&T Development Plan], http://www.most.gov.cn/tztg/200610/t20061031_ 37721.htm.
18. M. Cai, "'人造太阳'人类能寄予多大的厚望?" ["Artificial Sun": How great expectations can humans have?], Xinhuanet, 26 November 2006, http://news.xinhuanet.com/focus/2006-11/27/content_5366726.htm.
19. D. Clery, "ITER's $12 Billion Gamble," *Science*, 314 (2006), pp. 238–242.
20. Center for Space Thermal Science of Shandong University, http://219.231.136.132/home/Achievements.php.
21. National Bureau of Statistics of China, *China Statistical Yearbook 2006* (Beijing: China Statistics Press), p. 264.
22. Central Committee of Chinese Communist Party, "中华人民共和国国民经济和社会发展第十一个五年规划纲要" [Text of the Eleventh Five-Year Plan for the development of the economy and society of the PRC], 16 March 2006, http://news.xinhuanet.com/misc/2006-03/16/content_4309517.htm.

23. *China Statistical Yearbook 2006*, p. 218.
24. Ministry of Science and Technology, “关于开展创新型企业试点工作的通知” [Notice on launch of the work regarding the pilot project on developing innovative enterprises], 13 April 2006, http://www.most.gov.cn/tztg/200604/t20060426_31862.htm.
25. State Council of PRC, “中国洁净煤技术 ‘九五’计划和 2010 年发展纲要” [China’s Ninth Five-Year Plan and development outline for clean coal technology to 2010], 2003 http://www.coal.com.cn/CoalNews/ArticleDisplay_60263.html.
26. National Development and Reform Commission, “能源产业结构调整指导目录” [Guidance for restructuring the energy industry], 29 December 2005, http://www.ndrc.gov.cn/nyjt/nyzywx/t20051229_55159.htm.
27. 中国电力统计年鉴 2004 [*China Electricity Yearbook 2004*], p. 666; *China Electricity Yearbook 2006*, p. 473; 北京：中国统计出版社 [Beijing: China Statistics Press, 2004, 2006].
28. Organization for Economic Cooperation and Development, *IEA Electricity Information 2005* (Paris: OECD, 2005).
29. 张斌, 倪维斗, 李政 “火电厂和 IGCC 及煤气化 SOFC 混合循环减排 CO2 的分析.” <<煤炭转化>> 2005 年 01 期 [B. Zhang, W.D. Ni, and Z. Li, “Analysis of Conventional Power Plant, IGCC and Coal Gasification SOFC Hybrid with CO2 Mitigation,” Coal Conversion], no. 28 (2005) pp. 1–7.
30. Ministry of Science and Technology, “Enhanced Coal Gasification Technology,” *China Science and Technology Newsletter*, No. 440, 30 May 2006, http://www.most.gov.cn/eng/newsletters/2006/200606/t20060606_33788.htm.
31. “国务院原则通过核电中长期发展规划” [State Council passes the nuclear power medium- and long-term development plan], http://np.chinapower.com.cn/article/1001/art1001348.asp.
32. M. Dickie, D. Pilling, P. Hollinger, and M. Arnold, “Westinghouse wins $8bn China nuclear deal,” *Financial Times*, 18 December 2006, p. 4.
33. Ibid.
34. F. Jing, “Renewable energy gets huge outlay,” *China Daily*, 11 August 2005.
35. Renewable Energy Policy Network for 21st Century (REN21), “Renewables Global Status Report 2006 Update,” (Paris: REN21 Secretariat and Washington, DC: Worldwatch Institute), pp. 2–4, http://www.bmu.de/files/erneuerbare_energien/downloads/application/pdf/statusbericht_ee_en.pdf#search=%22Renewables%20Global%20Status%20Report%202006.%20Update%22.
36. “中共中央国务院关于推进社会主义新农村建设的若干意见” [Central Committee of Chinese Communist Party and the State Council of the People’s Republic of China regarding several suggestions on pushing forward the building of the new socialist countryside], 31 December 2005, http://www.gov.cn/jrzg/2006-02/21/content_205958.htm.
37. See Xinjiang Jinfeng Co website, http://www.goldwind.com.cn.

38. Securities and Exchange Commission (SEC), "Report of Foreign Issuer," 22 November 2006, http://www.sec.gov/Archives/edgar/data/1342803/000114554906001637/0001145549-06-001637.txt.
39. United Kingdom Department of Trade and Industry, *R&D Scoreboard 2006*, p. 108, http://www.innovation.gov.uk/rd_scoreboard/.
40. Calculated based on data from *China Statistical Yearbook 2006*.

5

Biotechnology Research in China

A Personal Perspective

Jun Yu

5.1 Introduction

The true "awakening of a giant"[1] in science and technology (S&T) started in the late 1970s, following the demise in 1976 of three Chinese political giants—Zhou Enlai, Zhu De, and Mao Zedong, marking the moment in time when the political pyramid of the People's Republic of China was shaken in a fundamental way. Chinese history took a dramatic turn thereafter and China's second-generation leadership, headed by Deng Xiaoping, orchestrated the more than decade-long process of restoring and reforming the national college education system. I am one of the Chinese scientists who benefited greatly from this process, being admitted into the Department of Chemistry at Jilin University in 1978.[i]

The foundation of China's current advancement in science and technology was laid in those first years of reform and opening to the outside world. By the time I graduated from university, the government had decided to promote programs that sent college graduates abroad. After a series of examinations, I was selected by the China–U.S. Biochemistry Examination and Application (CUSBEA) program, and subsequently went to New York University Medical School in

[i] The Chemistry Department of Jilin University was among the best in the nation in the late 1970s, and still is today. I majored in biochemistry, which was defined as a "fringe discipline" in the Chinese educational system (its curriculum was a simple mixture of biology and chemistry courses) though at the time biochemistry and molecular biology were already common departments in most Western universities or their medical schools.

1984 to pursue a PhD. Together with many Chinese young people, I was surfing the first wave of students going abroad. It is this very generation of students who are today bringing back new ideas, new fields of research, and new technology from the United States, Japan, Canada, and many European countries.

There are so few significant examples of advancement in biotechnology in China from the 1950s until the 1980s that the artificial synthesis of the complete bovine insulin stands in a class of its own. Despite the fact that more than three research groups have claimed credit for the synthesis,[2] the work done in China has its unique merit; it took half a decade, coordinated multi-institutional collaboration,[ii] and brought the artificially synthesized and bioactive insulin into a pure crystal form. In addition to the desire to enhance scientific research per se, this particular historic effort was also politically motivated, based on the belief of the renowned communist Friedrich Engels that life is the existence form of protein structures.

Most of the truly indigenous biological research and biotechnology developments in China have been and still are either heavily resource-driven or based on demands of national needs. The best examples are the fields of palaeontology (as an example of basic research) and hybrid rice breeding (as an example of applied research). The former relies on the Early Cambrian Chengjiang Fauna found in Eastern Yunnan Province and the Rehe Fauna discovered in Liaoning Province and Inner Mongolia, where the fossils of ancient animals and plants lived some 125 million years ago. The latter leveraged on wild rice resources and breeding experiences and won Professor Yuan Longping the title of the "Father of Hybrid Rice." Research was already being done in the field of genomics in China on a limited scale before China formally joined the international consortium of the Human Genome Project (HGP) in 1999.[3]

5.2 National objectives of biotechnology in China

It is next to impossible to summarize the precise national objectives of biotechnology, which are officially defined in the government's

[ii] Collaboration took place between two CAS institutes, the Institute of Organic Chemistry and the Institute of Biochemistry, and two universities, Fudan University and Beijing University.

five-year plans and factored into the plans of different ministries and governmental bureaus. Categories are broken down into agriculture, healthcare, forestry, energy, and oceanography; the first two are the most important, concerning the basic essentials of life, food and medicine. As the Chinese economy develops, the emphasis has shifted from agriculture to medicine, in addition to other important biotechnology fields, such as environmental protection and alternative energy.

Biotechnology figures prominently in the "Medium- and Long-term S&T Development Plan," published in 2006. The plan specifies biotechnology as the first of eight frontier technologies and lists five directions that are elaborated upon in Figure 5.1.

Detailed guidelines in the plan about basic research essentially propose three programs. The first program concerns frontier issues in different scientific research fields. Biology again is the first of eight listed fields and includes, for example, functional genomics, epigenetics, and systems biology. The second program relates to research

Figure 5.1 Five focus areas of biotechnology.

Drug target discovery	Functional characterization of key and disease-related genes; drug target screening and validation
Animal and plant models and drug design	Analysis and integration of bio-information; drug design and metabolism; computer-assisted designs, syntheses, and screening of compound libraries based on combinatory chemistry
Gene manipulation and protein engineering	Chromosome structure and site-directed integration; design and manipulation of protein-coding genes; polypeptide chain modification; structure solving; scaled protein purification
Stem cell-based human tissue engineering	Therapeutic cloning; directional differentiation; *in vitro* construction of structural organs and production; construction and damage repair of complex organs with multiple cell types
New generational industrial biotechnology	Scaled screening of functional microbes; modification of bio-catalysts and industrial production; bio-conversion media and systems for industrial operation

Source: Outline of National Medium- and Long-term S&T Development Plan (2006–2020).

activities that meet national demands, and has ten core topics that address questions concerning human health, agriculture, and the environment. The last section in this chapter is the most telling one, where only four large basic research projects are listed: protein, quantum regulation, nanoscience, and development-reproduction; in other words, two of these are biology-related. Each project is also elaborated on in detail. For instance, in the section for Development and Reproduction Research, the research keys are propagation, differentiation, and regulation of stem cells; genesis, maturation, and fertilization of the germline cells; control mechanisms of embryo development; somatic cell differentiation and animal cloning; declining of human reproductive function and mechanisms of degenerating diseases; safety and ethics in assisting reproduction, and stem cell technology.[4]

Beyond what is written in governmental guidelines, developing a significant industrial sector for biotechnology and biopharmaceutics is very much a goal of the government. Many of China's neighboring countries and regions have prioritized biotechnology and invested significant amounts of money in biotechnology initiatives. With regard to establishing a pharmaceutics industry, China is not only lagging behind Japan, Singapore, and Hong Kong, but also India, a developing country like China.[5] China is still considering what to do in the field of biotechnology research generally, and has not yet even begun to contemplate what to do in the field of application of traditional pharmaceutics and biopharmaceutics. At the very least, we need to see an effort to coordinate funding, funding mechanisms, and many organizations that host capable talents and are responsible for future biotechnology development.

5.2.1 Funding

The most significant funding institutions for biotechnology development in China are national (Ministry of Science and Technology (MOST), National Development and Reform Commission (NDRC), Chinese Academy of Sciences (CAS), and National Natural Science Foundation (NNSF); other ministries are also responsible in part, and the most notable ones are the Ministries of Health (MOH), Education (MOE), and Agriculture (MOA)). They support both basic research and technology development, while the rest of the funding agencies emphasize the needs of the economy, both at the local and national level. Accurate data on the total amount of government investment in biotechnology does not exist. My own estimate for 2004 is about 10–15 percent of the USD 3.7 billion collective budget. This figure is based on the total funding for major national programs for all fields of research and development. Though government expenditure on biotechnology research is bound to increase, the proportion of funds channeled into bioscience and biotechnology will not exceed 20 percent of the government's overall research and development (R&D) funding.[6]

Funding for biotechnology development has not been set aside any differently from other fields of science. If China is going to develop its biotechnology sector in the future, funding priority has to be loud and clear. Although MOST has a National Center for Biotechnology Development, it is equivalent to an administrative branch of the ministry and does not have special funds. In the United States, USD 51 billion was spent on biopharmaceutical R&D in 2005,[7] about thirteen times the amount spent in China. The annual national expenditure of China for biotech research is even smaller than the annual budget of one of Japan's major research institutes, RIKEN.[8]

The role of enterprises in biotechnology development has been very limited. Most of the Chinese pharmaceutical and biopharmaceutical companies are too small to contribute to the overall development of the industry. There are no equivalents to Pfizer (the top pharmaceutical company in the world) or Amgen (the top biopharmaceutical company in the world) on the horizon of China's pharmaceutical landscape. Most indigenous companies have focused their business on generic drugs and different formulas for traditional Chinese medicine. Innovative drug development is yet to play a major role in the Chinese pharmaceutical industry. There were only two Chinese pharmaceutical companies that made Standard and Poor's list rating of the top 100 corporations in China by sales in 2005 (sixty-third and ninety-sixth). These two had collective sales of about USD 2.3 billion.[9] However, their capacities for developing modern medicine independently are heavily reliant on R&D funding that is most likely from external sources, such as governmental funding agencies.

The most obvious consequence of this lack of adequate funding is that it makes scientists more "political," forcing them to devote a lot of effort to lobbying for the funding of large projects together. Scientists in China tend to spend most of their time networking and recruiting potentially prestigious colleagues (often members of CAS) to join them in submitting a grant application or competing among themselves for who gets more money. The single most important reason is the fact that a single research grant is not even enough to support a minimal research group composed of a PI (Primary Investigator) and his or her technician (or even a graduate student). For instance, in 2005, the NNSF of China granted RMB 2259 million (USD 280 million) to 9111 individual-investigator-initiated research projects (resembling R01 grants of the National Institutes of Health (NIH) in the U.S.).[10] Each grant is about RMB 250,000 (USD 30,000) for a period of three years and each individual investigator can only apply for one such grant from one funding agency in the same time period, so that the total cost per annum is about RMB 83,000 (USD 10,000). Even if we ignore the indirect costs in Chinese institutions (varying a lot, from five percent to 20 percent), compared to an average NIH R01 grant of USD 250,000 in direct costs per year, there is at least a 24-fold difference between researchers in China and in U.S. public institutions, not to mention the fact that most researchers in the United States have more than one grant at a time. Therefore, Chinese researchers have to do much better, and they have to continuously apply for more grants in order to keep up with the world's leading researchers.

The President of NNSF of China, Chen Yiyu, publicly vowed in 2006 to double funding for basic research in the next few years.[11] Government officials, especially at MOST, as well as government documents, have recently claimed that the main innovation bodies should be industrial enterprises, not research entities. But in reality, the government or governmental entities still provide most, if not all, of R&D funding today. Since the top-ranking companies are either completely or partly owned by the state, most of enterprise R&D expenditure is from the government too. The time is ripe for something to happen, but actions are slow in coming since implementation of major policy changes and the legislation process lag behind.

One of the most important issues is how R&D investment within a for-profit company is taxed. Under the current laws, companies receive no tax breaks. Therefore, R&D investment by an enterprise (especially a privately owned enterprise) has not really been encouraged. It remains to be seen what kind of regulations and laws actually materialize from the 2006 guidelines of the 15-year S&T plan, which are supposed to provide fiscal incentives to enterprises to invest in R&D.[12] A law equivalent to, or even beyond the scope of the Bayh-Dole Act passed in the United States in the 1980s, is what China needs to legalize technology transfer from public institutions, supported by public funding agencies.[iii]

5.2.2 Legislation

Legal issues concerning bioscience and biotechnology can be divided into two categories. One includes laws, guidelines, and statutes to ensure the development and innovation of bioscience and biotechnology, and the other deals with ethical issues provoked by bioscience and biotechnology. China's basic law providing guidelines for science and technology development was approved in 1993, and is no longer entirely compatible with the needs of society.[13] At present, improvements and changes to the law are being discussed. China also has guidelines to protect intellectual properties and natural resources, including genetic resources.

The MOST, MOH, and MOA, sometimes assisted by NNSF, are mainly responsible for the guidelines and statutes concerning biotechnology and bioethics. The guidelines provide basic legal guidance for the healthy development of bioscience and biotechnology in China. Most of these guidelines were drafted based on international conventions.

China's State Food and Drug Administration is currently undergoing intensive reform. It was originally founded on the basis of the State Drug Administration in the United States and is referred to as the Chinese FDA. The Chinese government's intention is to bridge the legal gap between China and the Western world.

5.2.2.1 Intellectual property rights and patents

Chinese scientists are gradually waking up to the importance of IP and patents. There are two single most important factors that hinder IP promotion in bioscience and biotechnology, or science and technology in general. First, China does not have a basic law on IP, an

[iii] The Bayh-Dole Act basically forms the foundation for U.S. research universities to enhance transfer of federally sponsored research to business and industry for public benefit. Prior to 1980, intellectual property (IP) derived from federal funding was to be claimed by the federal agency, and virtually no exclusive IP was licensed. Inventions developed rapidly and the biotech industry has blossomed since. The Bayh-Dole Act is possibly the most inspired piece of legislation to be enacted in the United States over the past half-century.

enforcement guideline, or a compatible system with rational mechanisms to implement the law and the guidelines.

Second, China needs a long-term national goal in bioscience and biotechnology development. For instance, although victory in the "War against Cancer," inaugurated by U.S. President Richard Nixon and Congress in 1971, has not yet been declared, the public money invested in this national project has been bearing fruit for the past 35 years. This money has not only benefited researchers and their institutions but also the pharmaceutical and biotech industries, as well as paved the way for a firm lead, at least for the Americans, in the push toward the Biotech Century. On the one hand, an accelerated and assured investment in the future industry from the government solicited private investment to convert intellectual property rights to merchandise. On the other hand, it assured the public support for IP-possessing as a new commercial initiative. The thriving biotech industry in the San Francisco Bay Area is the best testimony of such a brainy decision.

Though China does not have a basic law on IP, in principle, IP and patent protection are reasonably secured and protected in China by guidelines that are comparable to the world mainstream. In China, the dilemma is not whether there are laws and guidelines, rather the Achilles' heel in China is the enforcement of laws and guidelines; this will be an ongoing struggle for years to come. In my opinion, when the value of an investment is close to or equal to the value of IP and patents produced, we will see true protection of inventions because they can be valued in the market.

Patent applications have been increasing rapidly in recent years. The number of patents can be expected to continue to grow because the government has been emphasizing the importance of the ownership and generation of IPs, especially since China joined the World Trade Organization (WTO) in 2001. So far, the vast majority of patents applied for by Chinese researchers in the field of biotech and pharmaceuticals have not been so-called invention patents.

The government is largely responsible for both providing funds for IP development and for creating mechanisms to protect IP since the Chinese biotech industry does not generate much IP nor create many patents. The main reason is lack of funding to cover R&D as well as the application and maintenance of the patent. The same is true for many developing countries such as Brazil.[14] One solution would be that the fees incurred by the application and maintenance of a patent would be paid either by the funding agencies or the institutions (or companies) that provide institutional support for the research.

5.2.2.2 Ethics

There are many pressing ethical issues in the realm of biomedical or bioscience research, which need to be addressed carefully by scientists and society. These issues include privacy of personal genetic information and materials, human-embryo research, animal cloning, organ regeneration, stem-cell research, and genetically modified organisms (GMO). Since I am not an expert on all these issues, I am writing based on my personal experience and opinion concerning some of them.

Since the 1980s, Chinese policymakers have been putting into place a regulatory infrastructure to monitor biomedical activities. China is strongly opposed to reproductive human cloning.[15] However, human therapeutic cloning (wherein a patient's cells are used to form stem cells that grow tissues which may be needed by the patient) is legally permitted in China, which is different from the situation in many other countries, for example the United States and Germany.[16] For this reason, among others, China has been labeled a "country with probably the most liberal environment for embryo research in the world," and one where "some very sensitive embryo-based studies are conducted with little or no institutional review."[17]

I find these claims somewhat misleading. In 33 countries encompassing more than half of the world's population there is a permissive or flexible policy toward stem cell research.[18] Moreover, China has tried to do what is necessary to keep up with the rest of the world in terms of biomedical laws and guidelines. For instance, China's 1994 Law on Maternal and Infant Healthcare was not only disputed internationally but also within China. Chinese scientists also debated questions regarding how to handle problems arising from misunderstandings and unlawful behavior, such as how to prevent prenatal sex determination for abortions and screening for genetic diseases without informed consent. The law was the first of its kind in China. After a long period of discussion, a detailed guideline was issued by the State Council in 2001 to implement this particular law.[19] I am not trying to defend China and its many bureaucratic actions, but I believe that Chinese scientists and the Chinese people will change these misleading perceptions by open discussion and in forums among themselves and with the world outside China. "Rose-tinted glasses" are always there for some to use; but equally, not to use.

The best testimony for my statement is the fact that China was not only accepted as an active member of the international HGP Consortium in 1999, but is also involved in deoxyribonucleic acid (DNA) sample collections and sharing for the HapMap Project, the sequel to HGP, for studying genetic variations among human populations in different continents. Both the Chinese funding agencies and the institutions monitor research projects that involve human subjects. The cross-ethnic-group DNA samples collected by the scientists working on this project have been utilized efficiently for some other related projects.[20]

In addition to the principles relating to information-sharing, known as the Bermuda Principle, embraced by HGP Consortium, the consortium also has its Ethical, Legal, and Social Implication Committee to discuss and research issues arising from the project. In February 1997, at the second HGP meeting on large-scale sequencing strategy, held in Bermuda, HGP Consortium members and officials affirmed the principles of rapid, public release of genome sequence data, without restrictions on use. China, as a member of HGP Consortium, provided funding for the human genome research and released its data immediately.[21] Following the Consortium's principle, China and Chinese scientists have released the sequence data that they acquired from indica rice and the silkworm, as well as data from many public research projects.

The way to enforce the rules is to engage in active education and communication among ordinary people, scientists, relevant organizations, and law enforcement agencies; this can be organized and hosted by governmental organizations, the media or even at private gatherings. Again speaking from my own experiences, the consortium members of HGP, while actively working on their major projects, meet at least twice a year and communicate frequently online. Members pledge to follow the rules and the protocols, and sometimes an observer is dispatched to ensure that a witness who is not affiliated to a particular organization is present to defend unexpected allegations.

With regard to concerns about possible breaches of ethical behavior, one must remember that it will take time to make everyone aware of all the regulations and rules in all corners of this large country. The right attitude is one that promotes mutual understanding and professional interaction. This is the most practical course of action and eliminates hostile "saliva wars" or a war of words in a public debate. It is in the interest of Chinese scientists that implementation of regulations should be effective, for otherwise they will face difficulties in validating their research results internationally. According to traditional Chinese thinking, a friendly face-to-face dialogue among professionals is more easily accepted than public criticism. It is apparent that the lack of adequate discussions, especially those open to the public, is a weak link in the process of improving ethical standards. In April 2007 an initiative jointly organized by European and Chinese bioethicists and life scientists set up an expert group as part of the EU-funded BIONET project to develop a series of guidelines for the ethical governance of collaborative biomedical research between Europe and China. Open discussions between Chinese and foreign scientists are always an excellent path forward in solving conflicting opinions and misunderstandings in bioethics.[22]

Actually, since the 1980s China has seldom lagged far behind the international community when it comes to passing biotech regulatory guidelines. The present regulatory regime in relation to human embryonic and fetal stem cells, in general, "reflects emerging international standards."[23] In the bioethical field, China has been described as working toward "a recognizable liberal European framework of regulation, in several cases based on the British House of Lords Select Committee recommendations."[24] However, in contrast to the situation in Europe and in the United States where regulations have the force of law, in China they are, for the most part, guidelines and regulations without a basis in law. The question then arises as to what extent they are enforced. What many people, especially those who are not familiar with the Chinese culture and political system, do not understand is that it is much easier to pass a governmental guideline or statute than it is to enforce it. As a matter of fact, a lot of the guidelines are not enforced at all because the law enforcement is usually managed and interpreted by different governmental agencies under different ministries.

A recent series of publications, the Applied Ethics Series, sponsored by NNSF, lists and discusses most of the international declarations and almost all biotech laws, statutes, and guidelines in China.[25] The funding agency's decision to provide guidance for researchers and educational materials for the general public is a step in the right direction.

In the realm of GMO, China is ranked fifth in the world in terms of planting areas or production after the United States, Argentina, Brazil, and Canada. Although China has lagged far behind in developing and planting GMO, China has recently speeded up research activities in GMO, including cotton, rice, soy, corn, and rape seed. In fact, China's seed companies are currently competing with Monsanto, a global leader in GM seed production, for a dominant share of GM cotton in Chinese soil, where 25 percent of world cotton production takes place. Both parties own independent IP on genetically engineered genes that provide resistance to pests. However, laws and guidelines on GMO biosafety have not yet been established in China although the State Environment Protection Administration (SEPA) announced in 2005 that it was making efforts to this effect.[26] The National Biosafety Framework, drafted in the end of the 1990s, outlining the policy and regulative framework for biosafety management in China, is merely a guidance and not legally binding. In 2000, China signed the Cartagena Protocol on Biosafety[iv] to the Convention on Biological Diversity.[27]

Most Chinese people are actually not very religious, in the Western tradition of Judeo-Christianity, and the leaders of most policy-making bodies are not at all religious, so they should be able to think rather freely and objectively. Confucianism and Taoism have never been considered official religions; rather they are philosophies and wisdoms that have dominated the Chinese people's minds for centuries and are supposed to be functional and witty. Chinese cultural values do not appear to have favored basic research activities that foster systematic thinkers because no immediate benefits are expected from the investment. However, since the beginning of ancient civilization until the last dynasty, the Chinese were very good at craftwork and technology development, resulting in several technical advancements, such as the "Four Great Inventions of China."[v] Therefore, Chinese people are often practical thinkers, open-minded, solution-seeking, and capable of absorbing new ideas as long as they are thought to be beneficial for society, even if it is out of the ordinary in their thinking. For instance, GMO, animal cloning, and stem-cell research have not encountered resistance in Chinese society. This could, however, be due in part to the lack of public education and a public debate about these subjects, as well as due to the fact that these research areas have been emphasized for intensive development in the official S&T development plan, as I have discussed in the previous sections. For

[iv] The aim of the protocol is to regulate the international trade of GMOs.
[v] Papermaking, gunpowder, printing, and the compass.

the most part, researchers have been discussing ethical issues to do with GMO, animal cloning, and stem-cell research among themselves.

5.3 Biotech R&D institutions in China

Central or local governments, or governmental branches and ministries in general support most of the research institutions in China. However, the support is generally not enough to run a reasonable scientific research operation. The per-capita annual budget of these institutions ranges from RMB 100,000 to 400,000 (USD 13,000–52,000) and is usually composed of incomes or grants from both their own system (more like intramural grants of NIH in the United States) and other governmental funding agencies. The budgets of CAS institutes usually belong to the high end of the range, while those of university departments belong to the lower end. For instance, in 2004 an averagely funded institute, the Institute of Microbiology of CAS, had 307 research faculty and staff members and an annual budget of RMB 87,310,000 (USD 10.5 million) or RMB 284,397 (USD 34,000) per person. The annual average research budget for a research faculty member at the Institute of Genetics and Developmental Biology of CAS is RMB 312,875 or about USD 40,000.

5.3.1 Major institutions and some of their biotech spin-offs

It is easy to point out several top-notch research institutions and university departments from governmental research organizations, but not private ones. China is still a communist country where the People or the State representing them own most of the research institutions and universities. The largest institution in biological studies other than universities is CAS's Shanghai Institutes for Biological Sciences (SIBS). In addition, there are other biology-relevant academies and significant institutions that are managed by different ministries, such as the Chinese Academy of Agricultural Sciences of MOA, the Chinese Academy of Medical Sciences of MOH, and the Chinese Center for Disease Control of MOH.

5.3.1.1 The unique Shanghai Institutes for Biological Sciences

The SIBS is a unique organization within CAS; it is composed of six research institutions that focus their research activities on life sciences.

It is different from local branches of CAS in major cities, such as those in Shenyang, Chengdu, and Beijing. These branches include institutions for scientific research in other disciplines including biology, and they are managed independently. Among SIBS' six institutes and four associated research units, there are older institutes, such as the Institute of Biochemistry and Cell Biology, the Institute of Neuroscience and the Shanghai Institute of Materia Medica (which has its own independent legal status), as well as newly established ones, such as the Institute for Nutritional Sciences and CAS-MPG Partner Institute for Computational Biology. The SIBS has 1810 employees, including 1200 scientists and 30 academicians (gathered from both CAS and the Academy of Engineering) whose sheer number often serves as a measure of research capacity in China. It has an annual budget of nearly RMB 600 million (USD 78 million). Over 60 percent of the total budget comes from CAS and the rest is from grants from different funding governmental agencies.

Other than being the largest biological research institution in China, SIBS has two appealing features that are unique within CAS. First, it has foreign scientific directors in charge of several of its member institutes and research units. The second feature is that it has newly established research institutions and units to introduce new disciplines of biological research and to entice scientists back from abroad. Like other CAS institutions, SIBS is focused on basic research activities in the fields of biology. Drawing on the rich human resources in China's most cosmopolitan city and providing the most friendly living environment, even by international standards, anywhere in China, SIBS collectively is China's best research institution in biological research. But unfortunately it does not have any sizable biotech spin-offs.

5.3.1.2 The Beijing Institute of Genomics: From BGI to BIG

I am particularly proud of what my colleagues and I have achieved at the Beijing Genomics Institute (BGI), which also received praise from President Hu Jintao in his speech at the National Conference on Science and Technology in January 2006. He singled out seven landmark achievements in the history of science in China and three out of the seven are biology-related: hybrid rice, the synthesis of bovine insulin, and genomics research. The BGI, under CAS, has been directly involved in one of the seven listed achievements—genomics—and indirectly involved in hybrid rice research and developing high-performance computing technology. In the field of genomics, we are one of the half-dozen top genome centers in the world which are capable of sequencing, analyzing, and assembling large genomes, such as human and rice genomes.[28] The BGI has also sequenced the genomes of parental cultivars that give rise to a superhybrid rice.[29] The immediate goal of the project is to decipher the mechanisms of hybrid vigor at the molecular level. The BGI has been collaborating with the Institute of Computing Technology to develop high-performance computers because, in terms of immediate applications, the highest computing demand actually comes from the field of genomics and its sister field, bioinformatics.

The BGI is an interesting product of contemporary China. It started off as a research center within an institute of CAS (the former Institute of Genetics, and now the Institute of Genetics and Developmental Biology), was spun off to be a non-profit-making research institution (the Beijing Genomics Institute), and then was once again re-incorporated into CAS as a brand new institute (the Beijing Institute of Genomics, BIG). The BGI and BIG are now two separate entities; the former has about 450 employees and will be incorporated into a new biotech R&D company, while the latter, BIG of CAS with its one hundred or so faculty and staff, continues its mission to be a world-class public research institution in genomics. When BGI and BIG were sequencing large genomes, they had a combined annual expenditure of about RMB 140 million (USD 18 million) over a five-year period. Since the split, BIG has been fully supported by CAS with an annual budget of over RMB 20 million (USD 2.6 million) from CAS alone, and a similar amount of expected grants from other funding agencies.

5.3.1.3 Institute of Biophysics: From laboratory to market

The Institute of Biophysics is a typical research institute within CAS, focusing on basic research projects relevant to protein structures and functions as well as other programs on immunology, nanobiology, bioinformatics, and neural sciences.[30] The institute has over 60 professors as well as 500 graduate students. Its annual budget exceeds RMB 130 million (USD 17 million) and 91 percent of the total budget is from CAS and other governmental funding agencies. The institute's research groups have published several landmark papers in recent years. In addition, the institute founded two biotech companies, Zhongsheng Beikong Biotechnology and Science Inc. and Beijing Baiao Pharmaceuticals Co. The two companies had a total annual revenue of over RMB 60 million (USD 7.8 million) and over RMB 10 million (USD 1.3 million) profit in 2005.

The most successful enterprise founded by CAS institutions in the biopharmaceutical field is the Diao Group, spun off in 1988 from the Chengdu Institute of Biology of CAS. The Diao Group is focused on traditional herbs, biomedicine, and synthetic drugs; it has over 4000 employees and in 2005 had an annual revenue of RMB 1.2 billion (USD 155 million) as well as a comfortable profit (20 percent) of RMB 243 million (USD 31 million). There are also a few other biopharmaceutical companies, with products based on either Traditional Chinese Medicine (TCM) or biopharmaceuticals that were founded by universities and research institutions. One of the public companies is Tsinghua Unisplendour Guhan Bio-pharmaceutical Corporation Ltd, founded by Tsinghua University. Its principal activities are the research, development, manufacture, and sale of TCM products, Western medicinal preparations, biological medicaments, and health beverages. Its major products include Chinese herb medicines, glucose solutions, human serum albumin, and immune globulins. Its sales in 2005 were USD 271 million.[31]

Another example is China Bioway Biotech Group Co. Ltd., founded by Chen Zhangliang and Pan Aihua in 1992 when Chen was the Vice President of Beijing University. He is currently the President of the Chinese Agricultural University. He has a PhD from Washington University; yet another example of a returning student. China Bioway's business focuses on biotech products and biopharmaceuticals, including molecular diagnostics, vaccines, and recombinant proteins (such as insulin, human nerve growth factor, and human growth factor).

In general, universities are expanding their teaching programs to commercial applications of the life sciences, thus creating new educational options. In order to train life scientists for industries, a new program including 36 universities and colleges was initiated in 2002.[32]

5.3.2 Recruitment of overseas talent

Advancement in science and technology relies heavily on R&D support and training new talent. Aside from strengthening the current education system in China, a great resource is the overseas talent pool. China has been facing a brain drain since the early 1980s. There are several tens of thousands of college graduates each year seeking post-graduate education overseas. According to an estimate by a college professor from a highly ranked university, the top ten students in a newly graduated class of bachelor degree holders usually go abroad to pursue a higher degree, the next ten best students stay in their universities, and the rest either go to research institutes in China for post-graduate education or find jobs without continuing their education.[33]

Although a large number of Chinese students overseas are coming back each year, the majority are staying abroad. There are several reasons for this, based on my own experience. First, there is still not enough funding to support the returnees, even though an increase is expected in the future. The present personnel are grossly under funded. Institutions and universities hesitate to create new positions unless there are endowments and subsidies provided by funding agencies and endowment funds. Second, China's current economic situation is not strong enough to offer a salary close to what potential returnees earn if they stay abroad. A dramatic drop in financial security is too much of a threat. Third, the "soft environment" or social aspects of the living environment are not friendly enough in China. Young PhD scientists are most likely to have school-age children who may not be able to go to regular Chinese schools. The tuition fees of international schools in major Chinese cities are so high that they amount to three or four times a college professor's salary, not to mention the fact that they sometimes have more than one child.

Creating a friendly environment for overseas talent should be the highest priority, regardless of the difficulties. Educational institutions and commercial enterprises in China have finally realized this. As a result, new positions and biotech spin-off cultivating centers (often called incubators) are seen everywhere, mostly supported by local governments that offer both space and start-up (seed) money, although the sums are mostly rather nominal.

5.3.3 Publications

Publications are very important for Chinese scientists as the number of times they have published is often taken as an objective measure of career achievements, in conjunction with promotions. There are a small number of journals that are edited and published in China in the fields of biology (see Figure 5.2), but the quality of Chinese journals tends to be low. High-quality articles are sent to more prestigious international journals that have higher impact factors.[vi]

Broken down into disciplines, the papers published in international journals by Chinese scientists are concentrated in three major fields (35 percent of the total): chemistry (17,096), physics (11,606), and electro-engineering (10,227) in 2005. Biology ranked sixth after computer and material sciences. In domestic journals the top three fields were clinical studies, electro-engineering, and agricultural sciences. In this context, biology moved up to fifth place.[34]

Chinese scientists have published many high-quality bioscience and technology papers in recent years; some have been cover stories of the world's most prestigious journals, such as *Science, Nature,* and *Cell.* The sheer numbers of high-quality publications have been increasing each year.[35] Another trend is that a significant number of the high-quality publications are products of international collaboration.[36] International collaboration not only brings in expertise and resources to improve research quality but also contributes to high-standard publishing. A case in point are the back-to-back joint publications by international teams on the chicken genome in *Nature,* where both a genome sequence (genes for biological research) and a large amount of genetic variations (molecular markers for following up favorable inheritable traits) were provided to the research community and breeders.[37] Many major international scientific journals have started to invite prominent Chinese scientists to serve as co-editors. For instance, Chen Zhu who is the Director of Shanghai Institute of Hematology Human Genome Center and Ruijin Hospital, has been invited as co-editor of a prestigious journal—*Annual Review of Genomics and Human Genetics.* Wu Weihua (Chinese Agricultural University, College of Biological Sciences) and Lu Yingtang (College of Life Sciences, Wuhan University) are both the editors of *Plant Cell Report.*

[vi] A journal's impact factor (IF) is a measure of the frequency with which the "average article" in a journal has been cited in a particular year. The impact factor helps in evaluating a journal's relative importance, especially when compared to other journals in the same field. The impact factor for a journal is calculated by dividing the number of current citations to articles published in the two previous years by the total number of articles published in the two previous years.

Figure 5.2 China's leading biology-related journals.

Full journal title	IF	Est.	Sponsor	Publisher
细胞研究 Cell Research	2.161**	1990	Shanghai Institute of Cell Biology	Nature Publishing Group with SIBS, CAS
真菌多样性 Fungal Diversity	1.202**	1998	University of Hong Kong	Fungal Diversity Press
生物化学与生物物理学报 Acta Biochimica et Biophysica Sinica	0.505**	1999	Institute of Biochemistry and Cell Biology, SIBS	Blackwell Publishing
中国科学C类: 生命科学 Science in China Series C: Life Sciences	0.482**	1950	CAS and National Natural Science Foundation of China (NNSF)	Science Press
植物学报 Journal of Integrative Plant Biology*	0.432**	1952	Botanical Society of China and Inst. of Botany	Blackwell Publishing
生态学报 Acta Ecologica Sinica	1.414***	1981	Ecological Society of China	Science Press
植物生态学杂志 Journal of Plant Ecology	1.373***	1955	Institute of Botany, CAS and Botanical Society of China	Editorial office of Journal of Plant Ecology
中国应用生态学杂志 Chinese Journal of Applied Ecology	1.342***	1990	Chinese Society of Ecology and Inst. of Applied Ecology	Science Press
生物多样性 Biodiversity Science	1.295***	1993	Biodiversity Committee of CAS	Science Press
遗传学报 Acta Genetica Sinica	1.076***	1974	Genetics Society of China and Institute of Genetics and Developmental Biology	Science Press

Notes: * *Acta Botanica Sinica* was renamed Journal of Integrative Plant Biology in 2005, and the new journal has been indexed in SCI.

** Data from the Journal Citation Reports (2006) of ISI.

*** Data from the Chinese S&T Journal Citation Reports (2005) released by the Institute of Scientific and Technical Information of China.

Source: Data compiled by Wei Gong, Managing Editor of *Genomics, Proteomics, & Bioinformatics*.

5.3.4 International collaboration

Both the Chinese government and scientists value international collaboration, which has been very fruitful for scientific training and research. There are numerous special grants and scholarships for international projects and activities.

Permitting students and scholars to go abroad has been a policy of the government since the late 1970s. The most well-known and successful program for sending biology students overseas was the Sino-American overseas doctorate program, CUSBEA Program, started in 1981.[vii] Over 70 major U.S. universities joined the program initially and accepted graduate students from major Chinese universities. During the eight years of its duration, the program infused the United States with fresh blood, over 400 top graduate students.[viii] Two other programs, with similar motives, were CGP (Chemistry Graduate Program) for graduate students majoring in chemistry and CUSPEA (China-United States Physics Examination and Application) for graduate studies in physics. These programs were all initiated by China-friendly, prestigious American professors, organized on the Chinese side by top universities, and supported by the Commission of Education (now MOE). Over 900 students benefited from CUSPEA, initiated by Nobel Laureate T.D. Lee in 1979.

Due to the initial involvement in large international collaborative projects, China was also actively involved in the sequel to HGP, the International HapMap Project for mapping human population diversity. China took responsibility for nearly 10 percent of the project's work and provided 20 percent of the DNA samples from the Chinese population. Together with six other nations (United States, Canada,

[vii] The last character A in the abbreviation has been confusingly interpreted in three different ways: Administration, Admission, or Application, depending on the publication. All three interpretations may, however, be appropriate, depending on whom the letter A is geared toward administration for the program organizers, admission for the universities, and application for the students.

[viii] The first three groups of CUSBEA were successful in making it into academia in the United States, but the later groups were not as lucky. Many of them went into the pharmaceutical industry including start-up biotech companies in the mid-1990s. In 1995, at the peak time of one such pharmaceutical start-up, Human Genome Sciences, over a dozen CUSBEA students were working in the company of a few hundred employees.

United Kingdom, Japan, and Nigeria), China once again played an essential role in a worldwide collaborative research project. The significance of HapMap will become more obvious in the years to come when personal human genetic information is tailored to drug development and healthcare.

5.4 Current R&D activities in China

China has undoubtedly caught up in the "large-scale biology" sectors, such as genomics and proteomics, and is at a crossroads as to how to proceed to transform research achievements into commercial products. China has yet to seriously consider how to march across the "Valley of Death"[ix] between basic research/IP and development/ commercial products. Realistic road maps are desperately needed for all Chinese funding agencies.

5.4.1 Genomics

Genomics is the most noticeable field of all in bioscience. In the early 1990s, China did have its own initiative and plan to study the human genome, but it was not connected in any way to the original thoughts of a HGP and the international human genome research community. A sharp turn-about was made in 1999 when China joined the multinational HGP. Three national genome centers, one supported by CAS and two others supported by MOST were launched within a year. The centers funded by MOST are the Chinese National Human Genome Center in Beijing and Shanghai, headed by prominent Chinese biologists Qiang Boqin and Chen Zhu (academicians of CAS), respectively.

Chinese scientists at these genome centers are not only able to match the operational scale in sequencing large genomes, such as those of humans, rice, and silkworms, but they are also capable of developing computing tools independently for sequence assembly, gene finding, and annotations. There are only a handful of such genome research institutions around the globe which are comparable in scale and capability to these three Chinese institutes. If given

[ix] The "Valley of Death" refers to the time period prior to the demonstration of technical and economic feasibility of a new technological concept, when the risks are very high due to uncertainty or complexity.

enough funding and an operational mechanism to support a long-term technology development, these genome research institutes will allow Chinese scientists to forge ahead toward a new biological research era: the genomic era.

Peng Zhaohui and Gendicine

In this section, I will describe the process through which the world's first gene therapy drug came to the market in China.[38] Gendicine is the brand name of the gene therapy drug approved in October 2003 by China's State Food and Drug Administration (SFDA). Peng is the founder, the chairman, and the CEO of a biotech company, SiBiono Genetech Co. Ltd., which owns and produces Gendicine. It was founded in 1998 in Shenzhen half a year before a prototype of Gendicine; a normal human p53 gene engineered into an adenoviral vector was cleared for clinical trials (the p53 gene plays a key role in at least half of all human cancers when it is mutated). In May 2002, the company filed the first patent in China for the method to make recombinant adenovirus. The patent was granted in November 2004. Gendicine as a new drug was licensed in October 2003 and became the first marketed gene therapy drug in the world. The company has filed six domestic patents, three of which had been issued by mid-2006. It has also filed four patents for its drugs and methods internationally.

Although Peng and his colleagues do not have as much clinical experience as their American colleagues in developing genetically engineered therapeutic drugs,[39] they took an important step: to deliver a gene therapy drug directly to the market. The world's first clinical trial of gene therapy was performed in 1990 by R. Michael Blaese and W. French Anderson at NIH. In 1991 the Federal Drug Administration (FDA) in the United States published its guidelines for gene therapy in the United States and two years later clarified its regulations.

Similarly, Peng and his colleagues, together with the National Institute for the Control of Pharmaceutical and Biological Products, have helped China's SFDA put together a document, named "Points to Consider for Human Gene Therapy and Product Quality Control," governmental guidelines for gene therapy development in China.[40]

> The guidelines not only clearly state that gene therapy in China is restricted to somatic cells (not from reproductive organs), but also define protocols for quality control, safety and efficacy evaluations, and clinical trials of the gene therapy products.
>
> Determining the effectiveness of gene therapy is almost impossible as testing on humans is forbidden. Allegations that Gendicine has not been sufficiently tested are unjustified because all gene therapies that are based on protein-coding genes face the same problem—gene therapy has to be accompanied by protein therapy, in other words the patients have to receive both the gene (DNA) and its protein (often a genetically engineered product of the same gene for intravenous injection), which is functional. However, I do agree that there are cultural differences in attitudes toward lifesaving therapies and drugs. Chinese people traditionally expect some kind of a remedy from the doctor, especially doctors of TCM; this is especially true for less educated people who tend to be more willing to try new remedies.

5.4.2 Proteomics

The second promising field in Chinese biotechnology is proteomics. It has two basic directions. One is to study proteins in their spatial structures as mechanisms of operation (the basic tool is protein crystallography). The other is to address the question of how many protein components there are for a given cell type under a definable condition (the basic tool is mass spectrometry). The former leads to chemical definition of protein functions for drug discovery and design, and the latter yields information on molecular markers for disease diagnosis.

The Institute of Biophysics of CAS has been taking a firm lead in China in the study of proteins in their spatial structures as mechanisms of operation. Under Director Rao Zihe (now the President of Nankai University in Tianjin), who holds a PhD from the University of Cambridge, the institute's research groups have published several landmark papers in recent years. One example is the paper on the crystal structure of mitochondrial respiratory membrane protein complex II, and the other is a paper on the crystal structure of major light harvesting complex from spinach chloroplasts. Both mitochondrion and chloroplast are subcellular organelles that possess critical biological functions for animals and plants.[41]

The other direction in proteomics has been led by He Fuchu, the Vice President of the Academy of Military Medical Sciences. What differentiates He from other young scientists is that he does not have a doctorate from a foreign university. However, he has received excellent training in one of China's top universities, Fudan University in Shanghai. As a leading molecular biologist in China, He has organized several major projects interrogating gene functions. The most noteworthy project is China's Human Liver Proteome Project that aims to generate a comprehensive functional map of this vital organ, and which marks the genes and their relationship under physiological and pathological conditions. Many Chinese scientists are involved in the organization's research agenda. Protein research, as protein structure and proteomics research are loosely termed, was named one of the four major basic research programs in the government's Mid to Long-term S&T Plan.

5.4.3 Gene cloning

Gene cloning is a term for gene manipulation, also known as functional genomics. This is another field where Chinese scientists may make world-class breakthroughs, based on the sheer number of researchers involved, since the work in this field is hypothesis-driven, labor-intensive, and depends very much on creativity and capability. An excellent example is the work done by the current Vice President of CAS, Li Jiayang, an academician and veteran in plant molecular biology research. He has, over the past decade or so, published several influential manuscripts on genes, which he and his colleagues actually cloned that relate to plant mechanical property, architecture, and tillering.[42] Equally outstanding work has been carried out in the field of human genetics where the classical work is to clone genes whose defects cause diseases directly. The achievements in this field have been best represented by the work of several newly elected academicians of CAS such as He Lin, currently a Professor at Shanghai Jiaotong University heading a human genetics group studying neuropsychiatric diseases. He and his colleagues have successfully identified IHH as the disease gene for Brachydactyly A-1.[43]

Although there are still misunderstandings within the Chinese biological research community toward genomics, genomicists in China are taking a firm lead in the acquisition of genomic data and integration of genomic information. What has to be realized by Chinese biologists is the fact that genomics, especially DNA sequencing, is still key in paving the way toward systems biology, an emerging field that integrates basic genomic and biological information, acquired through diverse newly developed technology. Deciphering the genome sequence of an organism marks the very first step in a systematic study of its biology.

China does indeed have enough money to pursue genomic research and it has done so with very limited investment in the past few years. The molecular cloners who benefit the most are those who work in the research fields of human, rice, silkworm, and chicken because their fellow countrymen have sequenced the genomes, host the databases, and have a team exploiting the information in the genetic code of these important organisms. Gene cloning is the basic technology that paves the way for genetic diagnostics, creating genetically modified organisms, and gene therapy.

Genomics is, to a large extent, a scaled gene cloning process, and cloning is the process itself. When Chinese scientists are united in this conceptual frontier, China will march into the biological future. The reason is simple: the rest of the world is generating data and developing technology in this direction and China has to do so too, sooner or later. I envisage that China will continue taking the lead in genomics, proteomics, and functional gene cloning that benefits from the first two large-scale "sister fields."

5.4.4 Others

There are other fields of research worth observing in the future, such as stem-cell research, animal cloning, and developmental biology. One of the ultimate goals in modern biological research is to develop drugs to cure life-threatening diseases of humankind. It is regrettable that China has not been able to establish a strong pharmaceutical industry nor developed a noticeable number of its own modern medicines. The exception is artemisinin, a naturally occurring compound extracted from an active ingredient in sweet wormwood. It was initially isolated by Chinese scientists from traditional herb medicine used to treat malaria and hemorrhoids, especially for drug-resistant malaria largely due to its exceptional potency in killing the parasites through interference with the function of their mitochondria.[44]

Despite some of the concerted efforts for new drug development in China, it has not been realized that it is an expensive endeavor. Unless the Chinese government or private investors invest an adequate amount of money in the field, it is difficult to foresee an optimistic future for drug development.[45]

5.5 Level of R&D in China

The level of R&D in China, especially in the fields of biomedical research and technology development, can be measured in four different aspects: the areas of research focus, the distance of these areas from the frontiers, governmental expenditure on R&D, and the dynamics of the talent pool.

I believe that the strong areas in biotech R&D will prove to be the ones targeted in the government's "Medium- and Long-term S&T Development Plan." There are two reasons for this prediction. First, the areas for intensive future investment often have solid groundwork done previously, and second, research results will certainly follow when money is invested. The new R&D investment target is 2.5 percent of gross domestic product (GDP) in 2020. Despite the fact that the focus fields in the plan are spread out over frontier technology and basic sciences, the biology- or biotechnology-related ones are readily recognizable: protein science, drug development, new genetically engineered crops, and stem-cell research. These are not only the research areas that China has paid a lot of attention to in previous years, but are also some of the most important focal points of the world's S&T development for the future. China certainly has its chance.

There are many reasons why China can become strong in some of the major biotech areas, such as agriculture, health, energy, and the environment. One of them is that they have been targeted as nationwide priorities, which usually creates new opportunities and fosters new ventures. After all, the major investor in biotech development (or any frontier technology) is the government. Another important reason has to do with the recruitment of new talent and the introduction of new technology. Deng Xiaoping's opening policy has sown the seeds for over two decades. Most of the returnees are from the United States and the United Kingdom, where investments in biomedical research are significant. A great majority of Chinese students overseas have pursued PhD degrees in the fields of biomedical research; some of the pioneers have already worked in the United States for a couple of decades. In addition, technology is introduced to, and investments are made in the Chinese mainland from the overseas Chinese community at large, as well as from Hong Kong and Taiwan. However, the advancement of bioscience and biotechnology is not free of uncertainty. To a certain extent, progress will depend on what kinds of guiding principles and legislative assurances will be made in the new basic law on S&T development.

I do not have the slightest doubt that R&D funding in China will increase in the future. However, productivity may not be proportional to the money invested, and both misguided investments and mismanagement can lead to failures. First, China does not have strong legislative guidance to protect IP, while at the same time transferring them effectively to commercial entities for product development. Therefore, the "Valley of Death" for transferring biotechnology to commercial products is much longer and harder to cross in China than in many other countries.

Second, China simply does not have enough experienced experts in its talent pool. This is a conclusion reached on the basis of discussions with many colleagues who were educated abroad and have been working in China for half a decade or so. China does not have a sufficient number of experts who are visionaries with diverse talents, and who are capable of soliciting venture investment, steering research directions, and evaluating research programs. As a result, the value of research and the viability of business proposals are often not recognized and handled appropriately, while bureaucratic influence and manipulation are dominant. Similarly, professional organizations are not going to be strong enough to promote and discipline their members. As a result, true collaboration is bumpy, and research activities are often chaotic.

Third, governmental funding agencies have too little money and too much control. One of their major problems is a lack of standard management principles. Funding agencies always try to fund more people and projects whereas grant applicants always want to have more money. For instance, most of the research funds (although a small portion may be over-funded as a result of political maneuvering) are not realistic for carrying out real and efficient experiments, with the result that the same proposals have to be submitted to multiple funding agencies. Some undoubtedly receive token funding.

Fourth, the Chinese education system has some intrinsic problems for biomedical research and biotech development. A medical degree in China is only equivalent to a bachelor's degree in the United States and Europe, except at the Union Medical University in Beijing and a few specialized programs. Therefore, most Chinese medical doctors do not have the capability nor the interest to carry out research projects related to their medical professions.

Fifth, the Chinese education and research systems simply need a stable and long-term mechanism or standard for recruiting capable people, such as institutional tenure track positions and long-term endowments. For example, CAS's "100 Talent" Program has a period of three years only; it is not an endowment in the true sense but more or less a start-up fund.

Last, but not least, China needs a truly transparent and democratic decision-making process for the entire scientific research funding system. It may not necessarily change the political landscape at this point, but it would allow people who work within the system to know what is going on. Many of the research projects offered by the funding agencies are not well explained. Applicants are given few opportunities to discuss the project among collaborating parties. The participants of research projects are entitled to know who designs the projects and why. Otherwise, there are no responsible parties for a decision that spends taxpayers' money if the government is always granted immunity.

China has to find a way to build long-term mechanisms to unearth and cultivate its innovative power. One of the most important pieces of legislation is to set long-term goals and support research activities; the time frame for a project has to be longer than the tenure of the political leadership that set it in motion. In the current funding environment, a greater portion of the funding, such as the money from NDRC, has been devoted to building infrastructures, not to supporting research. Even infrastructure building constantly requires new injection of money since the life of high-tech equipment is limited to about three to five years.

5.6 Conclusion

China's fast growing economy has the means to provide sustainable support for its S&T development. China needs more talented people with visions that map out the path toward future S&T. China needs strong and large clusters of research groups with abilities to transform these visions into realities. China needs to build well-equipped and competitive research institutions and universities to host such talent. China needs adequate and relatively stable funding for the ambitious workforce. A strong biotech niche has to be created in several places, together with a friendly living macroenvironment that will attract domestic and overseas talent and their families.

We should be able to see major changes in China in the next couple of decades. Despite the challenges, one thing is absolutely certain: It will happen! We have to believe this when the world's most populous nation vows to bring it about. The Giant of the East is standing up to the world.

Notes

1. Z.Z. Li et al., "Health biotechnology in China—reawakening of a giant," *Nature Biotechnology*, 22 Supplement (December 2004), pp. DC13–18.
2. For instance, American scientist Christian B. Anfinsen who was credited with synthesizing and crystallizing this vital hormone shared the 1972 Nobel Prize in Chemistry with two other American scientists, S. Moore and W.H. Stein.
3. B.Q. Qiang, "Human Genome Research in China," *Journal of Molecular Medicine*, 82: 4 (2004), pp. 214–222.
4. State Council of the PRC, "国家中长期科学和技术发展规划纲要" [Outline of National Medium- and Long-term S&T Development Plan (2006–2020)], 9 February 2006, http://www.gov.cn/jrzg/2006-02/09/content_183787.htm.
5. For a comparison of Indian and Chinese pharmaceutical industries, see "Indian and Chinese Pharmaceutical Industries," ATIP06.019, 26 April 2006, http://www.atip.org/2006-Reports.
6. Ministry of Science and Technology, "The Annual Reports of the National High Technology R&D Program of China—the 863 Programs from 1986 to 2004," http://www. most.gov.cn/ndbg/index.htm.
7. "Biopharmaceutical Industry Research and Development Tops $51 Billion in 2005," Burrill & Company, 7 March 2006, http://www.burrillandco. com/burrill/pr_1141780182.

8. T. Takahashi, "Can China catch-up with Japan and Germany in 10 years," presentation at Conference "China's New Knowledge Systems and Their Global Interaction," Lund, Sweden, 29–30 September 2003, p. 12, http://c-faculty.chuo-u.ac.jp/~takumat/English/Research_Recommend_Books.html#.
9. The two companies are Shanghai Pharmaceutical (sixty-third) and Guangzhou Pharmaceutical (ninety-sixth). J. Bailey and X.M. Song, "Sizing up China's corporate elite," *BusinessWeek*, 29 July 2005.
10. National Natural Science Foundation, "国家自然科学基金资助项目统计资料 2005" [Statistics of Programs Supported by NNSF of China in 2005], http://www.nsfc.gov.cn/nsfc/cen/xmtj/psd/2005_table.pdf.
11. Y.Y. Chen, "发展科学基金制 推动自主创新 建设创新型国家" [Developing science funding system, promoting indigenous innovation, building innovative country], 科学时报 [Science Times], 23 May 2006.
12. Outline of National Medium- and Long-term S&T Development Plan (2006–2020).
13. Ministry of Science and Technology, "中华人民共和国科学技术进 步法" [A basic law providing guidelines for science and technology development of the PRC], 2 July 1993, http://www.most.gov.cn/flfg/fl/199307/t19930702_30637.htm.
14. M. Ferrer et al., "The Scientific Muscle of Brazil's Health Biotechnology," *Nature Biotechnology*, 22 Supplement (December 2004), pp. DC8–12.
15. K. Leggett, "China has tightened genetics regulation—rules ban human cloning. Moves could quiet critics of freewheeling research," *Asian Wall Street Journal*, 13 October 2003.
16. X.Z. Yang, "An Embryonic Nation," *Nature*, 428 (2004), pp. 210–212.
17. Ibid.
18. By permissive is meant that various embryonic stem-cell derivation techniques can be legally pursued including somatic cell nuclear transfer (SCNT), also called research or therapeutic cloning. The SCNT is the transfer of a cell nucleus from a somatic or body cell into an egg from which the nucleus has been removed. By flexible is meant stem cells derived from fertility clinic embryo donations only, excluding SCNT. See World Stem Cell Map in talk by W. Hoffman: "Stem Cells: Human Health, Global Competition and National Security," St. John's University and the College of St. Benedict, Collegeville, Minnesota, 19 November 2003, http:// mbbnet.umn.edu/scmap.html; http://www.heartacademy.org/newsletter/3/3.pdf.
19. D. Dickson, "China Brings in Regulations to Put a Stop to 'Genetic Piracy,'" *Nature*, 395 (1998), p. 5.
20. R.S. Spielman et al., "Common Genetic Variants Account for Differences in Gene Expression among Ethnic Groups," *Nature Genetics*, 39: 2 (February 2007), pp. 226–231.
21. China's share of the work was only one percent. This small share was due to work allocations that had already been decided and funded by the Consortium's early members. For more about the HGP see D.M. Muzny et al., "The DNA Sequence, Annotation And Analysis of Human Chromosome 3," *Nature*, 440 (2006), pp. 1194–1198.

22. Ethical Governance of Biological and Biomedical Research: Chinese–European Co-operation, http://www.bionet-china.org/.
23. O. Döring, "Bioethical standardisation: obstacles or catalyst of innovation in China," presentation at conference "Asia's Growing Importance in the Global Innovation System." European Alliance for Asian Studies, Hamburg, 19 March 2006.
24. Ibid.
25. 韩跃红 [Y.H. Han] (ed.), 护卫生命的尊严： 现代生物技术中伦理问题研究 [Standing up for the Dignity of Life: the Ethic Issues in Contemporary Biotechnology] 人民出版社 [Beijing: Renmin Chubanshe] 2005.
26. State Environmental Protection Administration, "我国将制定'转基因生物安全法'" [China will formulate a law on GMO biosafety], 20 May 2005, http://www.sepa.gov.cn/hjyw/200505/t20050520_66766.htm.
27. W.H. Yang, "Regulation of Genetically Modified Organisms in China," *Review of European Community & International Environmental Law (RECIEL)*, 12: 1 (April 2003), pp. 99–108.
28. J. Yu et al., "A Draft Sequence Assembly of the Rice (Oryza sativa ssp. indica) Genome," *Science*, 296 (2002), pp. 79–93; J. Yu et al., "The Genomes of Oryza sativa: A History of Duplications Rice genomes," *PLoS Biology*, 3 (2005), p. e38.
29. J. Yu et al., "A Comprehensive Crop Genome Research Project: the Superhybrid Rice Genome Project in China," *Philosophical Transactions of the Royal Society* (London: Royal Society, 2007).
30. Chinese Academy of Sciences, "中国科学院统计年鉴 (2005)" [*Statistical Yearbook of Chinese Academy of Sciences 2005*], 科学出版社，北京 [Beijing: Kexue Chubanshe].
31. Ibid.
32. Ministry of Education, "教育部、国家计委关于批准有关高校建立"国家生命科学与技术人才培养基地"的通知" [Ministry of Education and State Development and Planning Commission approving the national life science and technology talent culture medium bases for universities], 19 July 2002, http://www.moe.edu.cn/edoas/website18/info4470.htm.
33. Personal communication, July 2006.
34. Ministry of Science and Technology, China Science and Technology Newsletter, no. 423, 10 December 2005.
35. Z. Chen et al., "Life Sciences and Biotechnology in China," *Philosophical Transactions of the Royal Society*, B: Biological Sciences (London: Royal Society, 2007).
36. H.D. Chen et al., "Plant Biology Research Comes of Age in China," *The Plant Cell*, 18 (2006), pp. 2855–2864.
37. International Chicken Genome Sequencing Consortium, "Sequence and Comparative Analysis of the Chicken Genome Provide Unique Perspectives on Vertebrate Evolution," *Nature*, 432 (2004), pp. 695–716; International Chicken Polymorphism Map Consortium, "A Genetic Variation Map for Chicken with 2.8 Million Single-Nucleotide Polymorphisms," *Nature*, 432 (2004), pp. 717–722.

38. Author's discussions with Peng Zhaohui in 2006.
39. Z.H. Peng, "Current Status of Gendicine in China: Recombinant Human Ad-p53 Agent for Treatment of Cancers," *Human Gene Therapy*, 16 (2005), pp. 1016–1027.
40. "Points to Consider for Human Gene Therapy and Product Quality Control: State Food and Drug Administration of China," *BioPharm International*, 17: 5 (2004), pp. 73–76.
41. Z.F. Liu et al., "Crystal Structure of Spinach Major Light-Harvesting Complex at 2.72 Å Resolution," *Nature*, 428: 6980 (2004), pp. 287–292; Sun et al., "Crystal Structure of Mitochondrial Respiratory Membrane Protein Complex II," *Cell*, 121: 7 (2005), pp. 1043–1057.
42. Y.H. Li et al., "BRITTLE CULM 1, Which Encodes a Cobra-Like Protein, Affects the Mechanical Properties of Rice Plants," *Plant Cell*, 15: 9 (2003), pp. 2020–2031; X.Y. Li et al., "Control of Tillering in Rice," *Nature*, 422 (2003), pp. 618–621; Y. Dai et al., "Increased Expression of MAPKK7 Causes Deficiency in Polar Auxin Transport and Leads to Plant Architectural Abnormality in Arabidopsis," *Plant Cell*, 18: 2 (2006), pp. 308–320.
43. B. Gao et al., "Mutations in IHH, Encoding Indian Hedgehog, Cause Brachydactyly Type A-1," *Nature Genetics*, 28: 4 (2001), pp. 386–388.
44. Wei Li et al., "Yeast Model Uncovers Dual Roles of Mitochondria in the Action of Artemisinin," *PLoS Genetics*, 1: 3 (2005), p. e36.
45. M.W. Wang, "Biological Screening of Natural Products and Drug Innovation in China," *Philosophical Transactions of the Royal Society*, B: Biological Sciences, (London: Royal Society, 2007).

Acknowledgments

No book is written without the help of a wide array of people—anonymous reviewers, colleagues, friends, and family members. All the contributing authors are thankful to Teemu Naarajärvi, who has worked as the research assistant on this book project for more than 18 months. I, especially, am very grateful to him for his help and support. Among his many tasks, Teemu searched for data, took charge of the figures and compiled the index.

We would also like to thank Sitra, the Finnish Innovation Fund, for the financial support that made this book possible, and in particular Antti Hautamäki, Director of Innovation Research at Sitra, for devising the book's grand strategy.

I, for my part, would like to extend my gratitude to my five co-authors—Arthur Kroeber, Bai Chunli, Wang Chen, Jiang Kejun, and Yu Jun—for this unique multicultural team effort. They have all written several drafts and patiently answered my endless questions about both their drafts and high-tech research in China more generally. I am also thankful to my colleagues at the Finnish Institute of International Affairs for their continued encouragement and help during the book project. Mikael Siirilä who designed the more complicated figures for Chapter 1 deserves special mention.

Over 100 researchers and policy-makers working in the field of science and technology in China devoted their time to answering my in-depth research questions, which formed the basis of Chapter 1. These research interviews were done on the condition of anonymity, so I express my appreciation to them all collectively. Moreover, I would like to thank several people who have given me advice, provided expertise or commented on various parts of the manuscript during the course of this project: James Brock, Cao Cong, Eddie Chen, David Cowig, Ole Döring, Gu Shulin, Wolfgang Hennig, Scott Kennedy, Martin Kenney, Li Zheng, June Ling, Paul Mooney, Ou Longxin, Ren Hongxuan, Henry Rowen, Denis Simon, Richard P. Suttmeier, James Thomas, Halla Thorsteindottir, Kathleen Walsh, Wang Zhonglin, Xie Sishen, Xue Lan, Yang Fuqiang, Zhang Gang, Zhang Qiang, David Zweig.

Finally, I am indebted to my better-half, Chris Lanzit, who has not only encouraged me throughout this long process but has also tirelessly dug up relevant information about the topics of the book for me and my co-authors, as well as commented on several drafts of Chapter 1.

Beijing, 5 February 2007
Linda Jakobson

Index

(covers main text of Introduction and Chapters 1–5)